**Editor**
Lorin Klistoff, M.A.

**Managing Editor**
Karen Goldfluss, M.S. Ed.

**Illustrator**
Kelly McMahon

**Cover Artist**
Brenda DiAntonis

**Art Manager**
Kevin Barnes

**Art Director**
CJae Froshay

**Imaging**
Alfred Lau

**Publisher**
Mary D. Smith, M.S. Ed.

**Author**

*Bev Dunbar*

(Revised and rewritten by Teacher Created Resources, Inc.)

This edition published by *Teacher Created Resources, Inc.*
6421 Industry Way
Westminster, CA 92683
www.teachercreated.com

ISBN-1-4206-3525-5

©2005 Teacher Created Resources, Inc.

Made in U.S.A.

The classroom teacher may reproduce copies of materials in this book for classroom use only. The reproduction of any part for an entire school or school system is strictly prohibited. No part of this publication may be transmitted, stored, or recorded in any form without written permission from the publisher.

# Table of Contents

**Introduction** .................................................................. 3

**How to Use This Book** ..................................................... 4

**Revisiting 0–9** ............................................................... 5

    Activities ..................................................................... 6

    (Ideas for Counting by 1s–Ideas for Counting By 2s—Using Dot Cards—Using Digit, Number Word, and Tally Cards—Using 0–9 Spinners—Using Ordinals 1st–10th/First–Last—Check-Up 0–9)

**Exploring 10–90** ........................................................... 23

    Activities ................................................................... 24

    (Clap to 10—Counting by 10s—Pass the Ball—Skipping by 10s—What a Lot of Fingers—Reference Pictures—Make It 10—Plenty of Pasta—What Else Comes in Tens?—Using 10–90 Numeral/Finger/Number Word Cards—How Far Can You Go?—Paper People Chain—Sort the Lion's Mane—Which Path?—That's a Lot of Students!—Recording Tens—How Many Tens?—Mental Mind Munchers—Check-Up 10–90)

**Exploring 11–19** ........................................................... 49

    Activities ................................................................... 50

    (What's So Special About 11–19?—Making Patterns—Guess and Check—That's a Lot of Jumps—Let's Make Music—How Many Are There?—Bundling 10s and Extras—Hit the Target—How Many Can You Find?—Find a Number—Back to Front—Number Roll-Up x 2—Please Sit Down—Frog Jumps—Join the Dots—All Stations to Central—Paw Prints—Who Goes Where?—Which Position?—Roll-a-Snake—Mental Mind Munchers—Check-Up 11–19)

**Exploring 0–50 and Beyond** ............................................ 81

    Activities ................................................................... 82

    (Count With Me—What's Missing?—What's My Pattern?—Can You Make It?—Exchange to 50—Tally It—Guess How Many—Show Me 50—Croc Spots—Paw Prints—Mental Mind Munchers—Check-Up 0–50)

**Skills Record Sheet** ....................................................... 94

**Sample Weekly Program** ................................................ 95

**Blank Weekly Program** ................................................... 96

# Introduction

At last! Here are many carefully sequenced teaching ideas for developing numeration skills for numbers 0–50 with your students.

Making a teacher's life easier has been a major aim in this series. With the suggested activities, the activity pages, the sample plans, and the Skills Record Sheet (page 94), you will find planning your numeration program for the year to be more manageable.

A summary at the start of each unit identifies the key mathematical ideas. There are enough suggestions in each unit to have up to a whole class studying a math topic for at least a week! The Sample Weekly Program (page 95) shows you one way to do this.

Every activity is fun, easy to implement, and easy to understand. Each one is carefully designed to maximize the way in which your students build up their knowledge of our Base 10 counting system. Students are encouraged to think and work mathematically with an emphasis on mental recall and practical manipulation of objects.

A second book, *Math in Action: Operations Activities 0–50* (TCR 3526), covers simple addition, subtraction, multiplication, and division with whole numbers to 50. Together, they provide you with the practical number resources you need to keep both your young students and yourself keenly motivated. You will look forward to sharing the joys of exploring numbers to 50 with your students.

# How to Use This Book

## Activities

Many easy-to-use activities exploring numeration to 50 have been placed into four separate units to enable you to readily plan and implement activities for the whole class, groups, and individuals. You will never again run out of ideas about what to teach your students!

Start by revisiting the numbers 0–9 and then look at groups of 10 from 10–90 from as many perspectives as possible. Next, explore numbers 11–19 where you will see each number as a group of 10 and extras. Build on these place value understandings in the final unit by exploring numbers from 21–50 and beyond.

Each unit includes pages for activity cards, games, and practical activity pages, as well as an end-of-unit check-up. The activities, within each unit, can generally be adapted to the study of any number. Remember to revisit a favorite activity later in the year using higher numbers. The objective is to develop knowledge, skills, and understanding for the numbers 0–50 in a variety of fun, child-centered ways. The overall outcome for each unit is to estimate, count, compare, order, and represent whole numbers up to 50, with a special emphasis on the development of place value and mental recall.

Each activity includes coded specific skills to help your planning, implementing, and unit assessment. Look at the codes below.

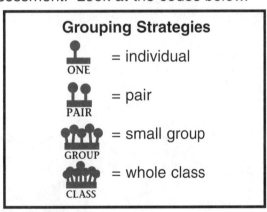

## Skills Record Sheet

A complete list of skills is provided on page 94 to help you keep record of how and when these outcomes have been reached.

## Sample Program

A special feature of this book is the Sample Weekly Program on page 95. You can see one way to organize a selection of activities from the "Exploring 10–90" unit as a five-day unit. A blank weekly program sheet is included on page 96 for your individual use.

# Revisiting 0-9

**In this unit, your students will:**

- ❑ Count forward/backward by 1s and 2s
- ❑ Identify a range of number patterns
- ❑ Identify groups as odd or even
- ❑ Estimate, model, and count 0–9 objects
- ❑ Match and order sets/numerals/words 0–9
- ❑ Use, read, and order "1st" to "10th" or "first" to "tenth"
- ❑ Record 0–9 as numerals, words, or tally marks

# Activities

### Ideas for Counting by 1s

- ❑ Have each student take a large handful of counters. Have each student guess how many each student has. Check by counting.
- ❑ Have each student turn over three or more dominoes. Have each student count how many dots altogether. Guess first and then check.
- ❑ Ask someone to call out a number that is larger than 10 (e.g., 17). Start counting backwards together from that number. Try starting from numbers larger than 20, 30, 40, or even 50.
- ❑ Have students count how many of the following: slices of bread in a loaf, pages in a favorite story book, buttons on a shirt, trees on the playground, whiskers on a cat, or petals on a flower.
- ❑ Give every student a small box of raisins (or candies). Have them count how many in their boxes, then count backwards as they eat each one! Ask, "What is the largest number of raisins (or candies) anyone has counted?"
- ❑ Have each student count the number of times someone can do an action. For example, how many times someone can jump, without stopping, a skipping rope held by two friends or bounce a ball without stopping?
- ❑ Tally up how many pets are owned by each of the students. Count how many altogether.
- ❑ Count forward or backward with your eyes shut, with your heads between your legs, or lying on the ground.
- ❑ Find a place on the playground where students can watch the traffic. Have them count how many vehicles go past within a time limit (e.g., 1 minute).
- ❑ Have students count how many mouthfuls of water it takes to drink a cupful.
- ❑ Using a calendar, have students count how many days until a given student's birthday.

### Ideas for Counting by 2s

- ❑ Have students count how many socks are in the whole class.
- ❑ Have students count body parts, such as eyes, ears, arms, hands, knees, legs, and nostrils. Count forward and backward by 2s.
- ❑ Have each student take a large handful of counters. Then have them guess how many each person has and check by counting by 2s.
- ❑ Challenge the class to find the student who can count to the largest number by 2s.
- ❑ Count the class by 2s as they come in from recess or lunch.

Revisiting 0–9

# Activities

## Using Dot Cards

### What to Do

- Make a copy of the Twos Pattern (page 8) and the Random Pattern (page 9) 0–9 Dot Cards. Laminate, and then cut out as sets of ten cards for whole class demonstrations.
- Make a copy of the Twos Pattern and the Random Pattern 0–9 Dot Cards for each group (e.g., yellow paper for "twos," and green for "random"). Cut out along the dashed lines as sets of ten separate cards.

### Whole Class Ideas

- Use a mixture of random and pattern cards. Shuffle the large cards. Hold up one card for four seconds and then hide it again. Ask students, "Guess how many dots you saw? (e.g., whisper your answer to the person sitting next to you). How do you know?" Discuss strategies. Check guesses by counting. Repeat this activity, gradually making the viewing time shorter and shorter. Do the guesses get closer? What is the shortest viewing time you need? Are some cards more difficult to guess closely?
- Shuffle the cards. Ask individual students to sort them into counting order forward or backward, according to the number of dots.
- Hold up three cards at random. Ask students to identify which one has the most dots. Reinforce spatial language—e.g., "the one on the right," "the one in the middle," or "the one on the left."
- Hold up a card at random. Everyone races to form a group with the matching number of students in it.
- Hold up a random card. Ask students to hold up fingers to match the number of dots. Repeat, but this time individuals race to collect a matching number of objects.
- Hold up a card at random. In pairs, create a story using that number to tell to another pair (e.g., 7—"My mom paid seven dollars for my lunch yesterday.").

### Small Group Variations

- Model the whole class actions using the smaller cards.
- Give each student nine counters. Rearrange the counters to match the twos pattern dot cards. How many other patterns can they discover using the counters?

Twos Pattern 0–9 Dot Cards  Revisiting 0–9

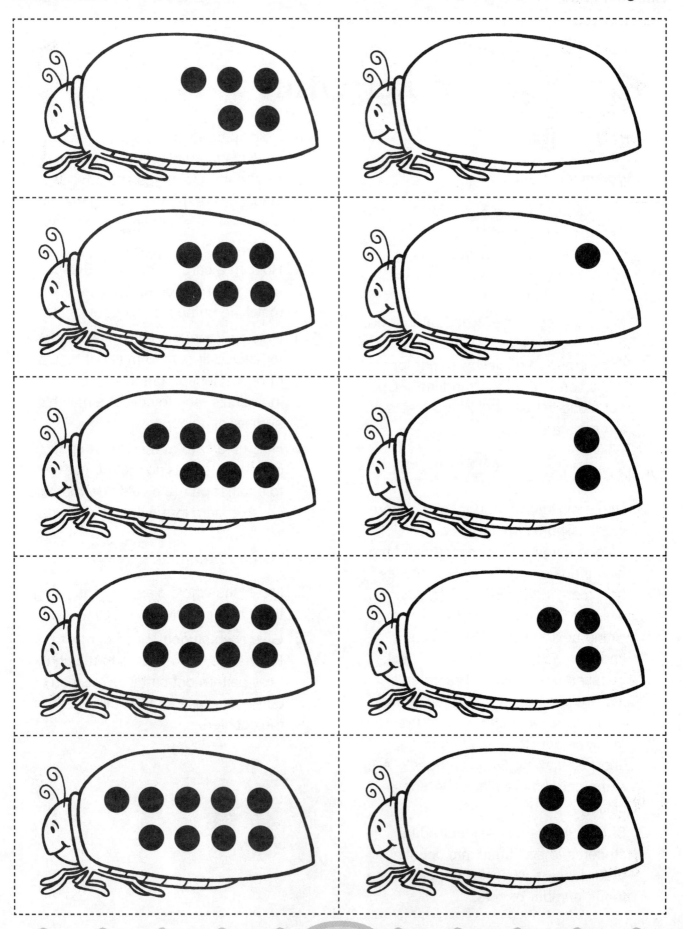

Twos Pattern 0–9 Dot Cards  Revisiting 0–9

**Random Pattern 0–9 Dot Cards**  Revisiting 0–9

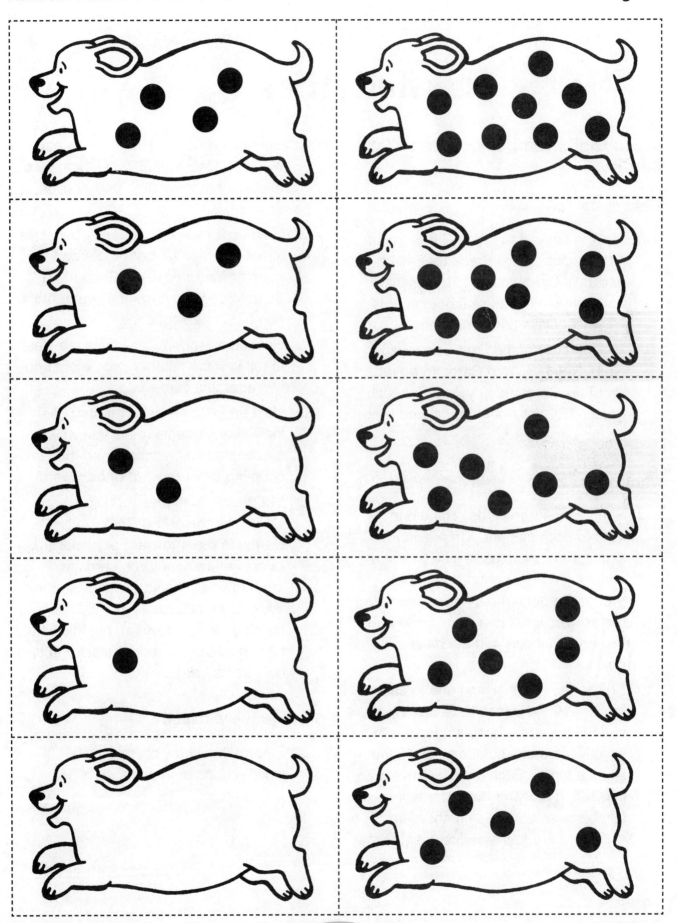

Random Pattern 0–9 Dot Cards  Revisiting 0–9

# Activities

## Using Digit, Number Word, and Tally Cards

### What to Do

❏ Make a copy of the Digit Cards (page 11), Number Word Cards (page 12) and Tally Cards (page 13). Laminate, and then cut out as sets of ten cards for whole class demonstrations.

❏ Make a copy of the Digit Cards (e.g., blue), Number Word Cards (e.g., red), and Tally Cards (e.g., yellow) for each group. Cut out along the dashed lines as individual cards.

### Whole Class Ideas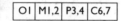

❏ Shuffle the large cards. Hand one card to each student. On a given signal, have them each race to find a matching number of objects. Form groups. In each group, have them put themselves into order from those with the least to those with the most objects.

❏ Shuffle the cards. Hand one to each student. On a signal, have them race to find partners—those students with exactly the same numbers. Have them sit as a team. Each team will invent an interesting story, or think of a fact to tell the class about its number (e.g., 4: "There are four sides on a square.").

❏ Shuffle the cards. Hand one to each student. On a signal, have them race to sort themselves into odd or even groups.

❏ Hold up a random card. Ask the class to tell you the number before and the number after this card. Encourage them to develop a mental image of the numbers in a sequence.

❏ Hold up a random card. Ask the class to tell you the number two before and two after this card.

❏ Hold up three cards at random. Ask individual students to tell you which one comes first, second, and third in counting order forward or backward.

❏ Put the cards in one set (Digit Cards/Number Word Cards/Tally Cards) in order forward or backward. Secretly turn over three cards at random while the students close their eyes. Can individuals identify the missing cards? Repeat, but this time mix up all the numbers before you turn over some cards.

### Small Group Variation

❏ Repeat the whole class activities in small groups or in pairs.

**Digit Cards**          Revisiting 0–9

| 0 | 1 | 0 | 1 |
|---|---|---|---|
| 2 | 3 | 2 | 3 |
| 4 | 5 | 4 | 5 |
| 6 | 7 | 6 | 7 |
| 8 | 9 | 8 | 9 |

Number Word Cards — Revisiting 0–9

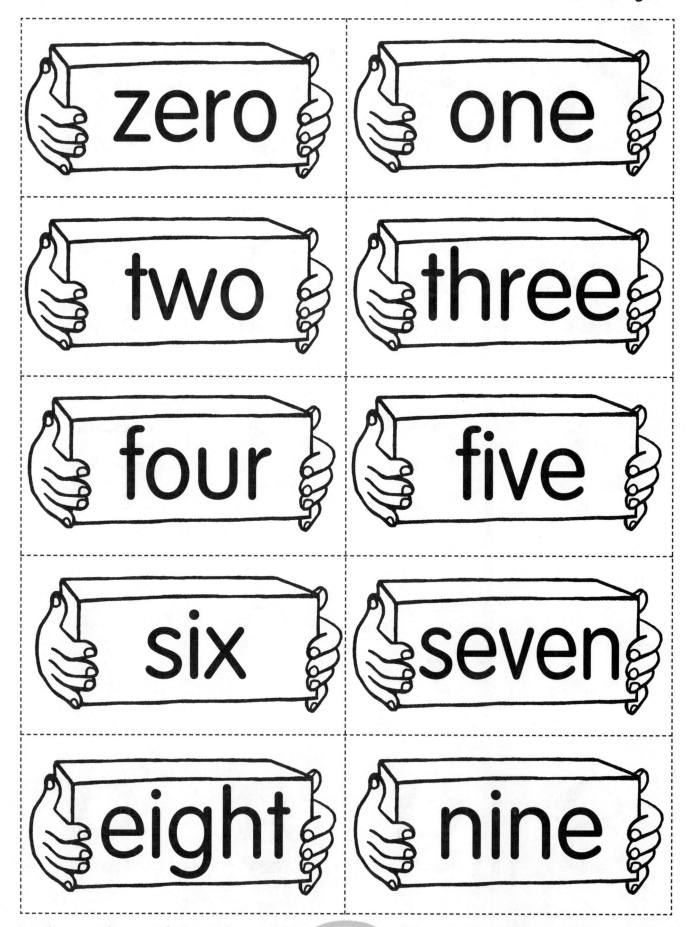

**Tally Cards**  Revisiting 0–9

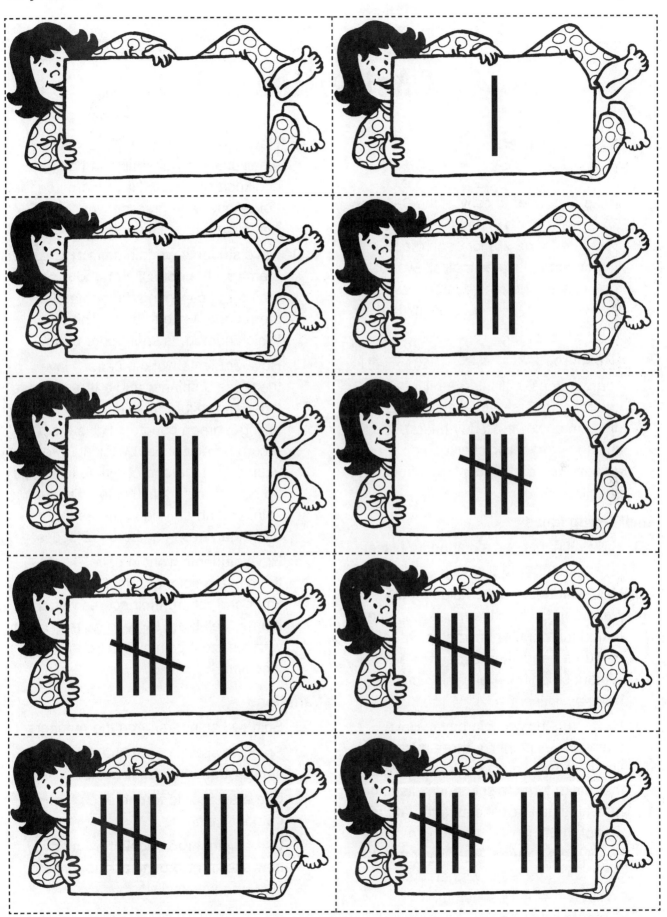

Tally Cards  Revisiting 0–9

# Activities

## Using 0–9 Spinners

### What to Do

- Make a cardboard copy of the 0–9 Dot Spinners (page 15) and 0–9 Numeral Spinners (page 16) for each group. Cut out and place a small skewer or toothpick through the center. Test each spinner for any bias (i.e., check that it lands at random).

### Other Resources

  M1,2 | R1,2,3 | P4

- board games, balls, three-minute timers, counters (e.g., plastic frogs, dinosaurs, clowns, teddy bears), chalkboards, chalk, chalkboard erasers, red/green counters, paper, pencils

### Small Group Ideas

- Have students use a spinner in place of dice in a board game.
- Have students work outside on the playground. Have them use a spinner to find the number of times a student can do a handstand or bounce a ball without stopping. Have them create more activities.
- Have students work in pairs. Have them use a spinner, a three-minute timer, and a pile of counters. Have them take turns to spin and collect the matching number of counters. At the end of the time limit, have them count up all the counters. The player with the most counters is the winner. Ask, "What is the largest number of counters collected?"
- Have students work in pairs with a small chalkboard, chalk, and a chalkboard eraser. Have them use the dot spinner and practice writing matching numerals on the chalkboard.
- Have students work in pairs. Have them use the dot spinner and record their turns by writing the numeral or the number word to match the dots. Which number is spun more often?
- Have students work in pairs. Have them use a spinner and a three-minute timer and record an even score by taking a green counter. Have them record an odd score by taking a red counter. At the end of a time limit, add up how many of each color. Did the spinner land on odd numbers more often than even numbers? Discuss.
- Have students work in pairs. Have them use a spinner and a three-minute timer and record their scores with tally marks. Tell them to try to be the player with the most points at the end of the time limit.

### Variations

- Spin two spinners. Add the numbers.
- Spin two spinners. Subtract the smaller from the larger number.
- Use a spinner to form groups for an outdoor game.
- Use a spinner to decide how many new spelling words to add to each student's weekly list!

0–9 Dot Spinners                                    Revisiting 0–9

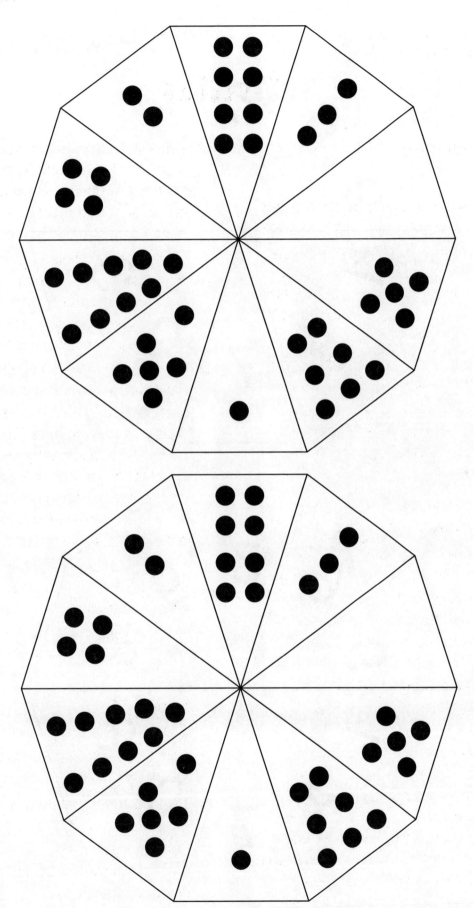

15

©Teacher Created Resources, Inc.                    #3525 Math in Action

# 0–9 Numeral Spinners

## Revisiting 0–9

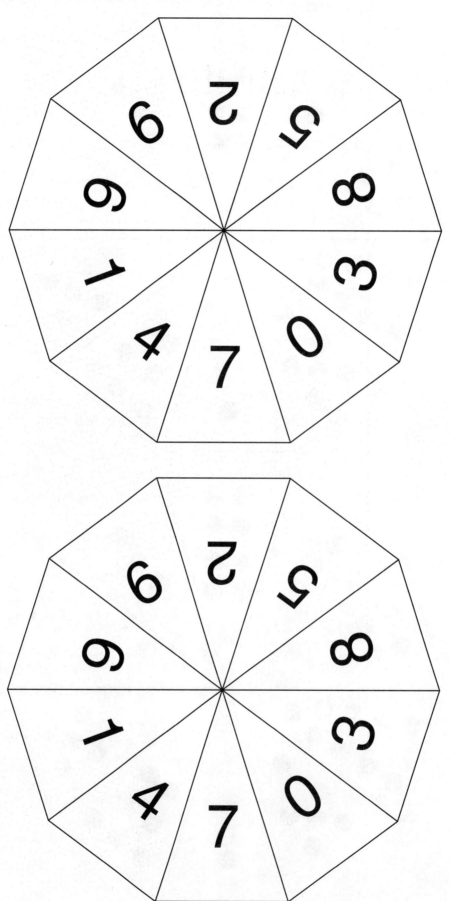

#3525 Math in Action 16 ©Teacher Created Resources, Inc.

# Activities

## Using Ordinals 1st–10th/First–Last

### What to Do

- Make a copy of the 10 figures (page 18) for each group. Color and cut out as 10 rectangular cards.
- Make a copy of the ordinal cards "1st–10th" (page 19 on bright blue paper) and "first-last" (page 20 on bright green paper) for each group. Cut out as sets of 10 cards.

### Other Resources

- 10 objects in different lengths or heights for each group (e.g., wool, sticks, leaves, containers)

### Small Group Ideas

- Have students sort the 10 cards into sets. Ask, "How many different ways can you sort them? (e.g., boys/girls, happy/sad, hands folded/not folded). Describe your sorting criteria. How many in each group? Is there the same number in each pile?"
- Have students create a pattern using at least six figures in a line. Have them ask each other to discover and then describe the pattern and to predict the next few figures.
- Have students line up the 10 figures and match with an ordinal numeral card and an ordinal word card. Have them sort in a line from left to right, or top to bottom. Ask questions using position language (e.g., "What color is the fourth person?" "Describe the one after the seventh figure." "What color is the one between the third and the fifth figure?").
- Have students work in pairs. Have them tell their partners how they would like each of the figures placed (e.g., "Put the boy in the yellow first. Make the girl with the red skirt second. Make the last one sad.").

### Whole Class Variations

- Each group races to place 10 different objects into length order, and then place the matching ordinal numeral and word cards under each object. Once checked, have them mix up the materials and then move to the next group. Repeat.
- Ten students stand at the front of the class. Ask individuals to call out instructions to place the students into height order (e.g., "Angelo, you go first.").
- Ten students stand at the front of the class. Hand 10 seated students the "1st–10th" cards. On a given signal, these students race to stand in front of the persons matching their cards. Have students check and then repeat using the "first"–"last" cards.

# 10 Figures

**Revisiting 0–9**

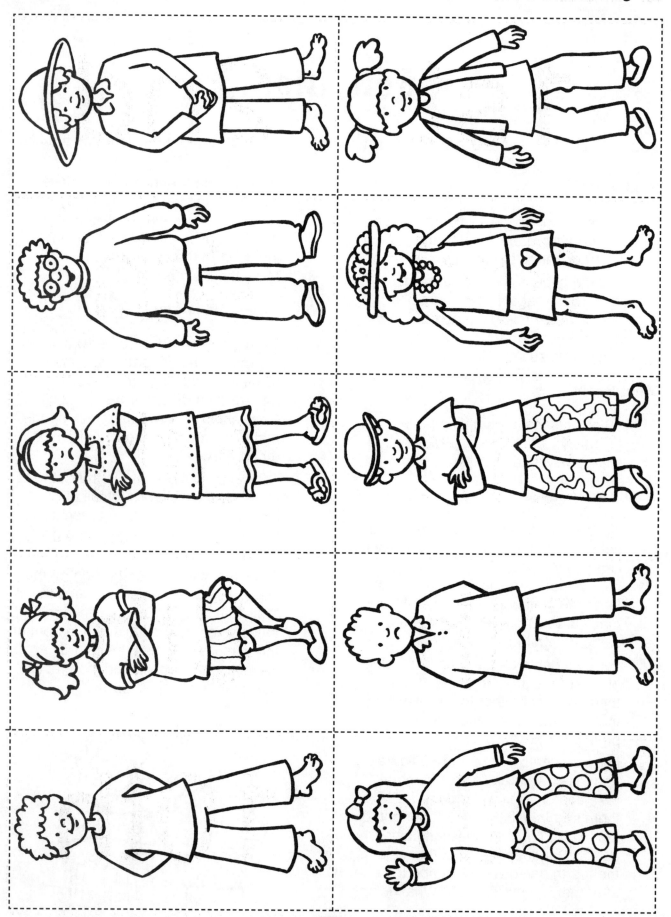

1st–10th Ordinal Cards

| | |
|---|---|
| # 1st | # 2nd |
| # 3rd | # 4th |
| # 5th | # 6th |
| # 7th | # 8th |
| # 9th | # 10th |

First–Tenth Ordinal Cards

Revisiting 0–9

| first | second |
|---|---|
| third | fourth |
| fifth | sixth |
| seventh | eighth |
| ninth | tenth |

# Check-Up 0–9

**Resource**
Check-Up 0–9
(page 22)

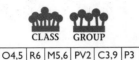

| O4,5 | R6 | M5,6 | PV2 | C3,9 | P3 |

**Activity**
- Use this worksheet to assess each student's competancy with the numbers 0–9. Teachers' comments have been added for your use. Feel free to use these as suits your needs.
- Record students' responses on the Skills Record Sheet (page 94)). The following are suggested instructions.

**Patterning**
- Tell students, "Look at the bead pattern at the top of the page. Continue the pattern by drawing three more beads. Write the number that tells you how many beads altogether. Put a circle around the sixth bead. Put a cross under the second to the last bead."

Sample Profile Comments
- "Luke can continue a pattern."

**Ordering**
- Tell students to look at the "Join the Dots" picture. Tell them to start at 9 and then join all of the dots in order backward. Tell them to write what they think they have drawn! Ask, "How will you finish your picture?" *(Join the 0 to the 9.)*

Sample Profile Comments
- "Although Abi can sort the numerals 0–9 into order forward, he still needs assistance to do this backward."

**Counting**
- Tell students to look at the four butterflies and to look at the number written on each body.
- Ask, "On every butterfly, write the number that comes before each number on the left wing. Write the number that comes after this number on the right wing."

Sample Profile Comments
- "Kay can identify and write one more and one less than any number to nine."
- "Most number formations are correct. He reverses '6' but is working on it!"

**Matching**
- Tell students to look at the smiley faces under the butterflies. Say, "Write the number word *six* in the space to the left of the first face. Count how many faces. Now draw more faces so that there are six faces altogether."

Sample Profile Comments
- "Karin has developed a strategy for solving simple addition problems."

**Ordering**
- Tell students to look at the ants at the bottom of the page. Say, "Draw lines to match the cards to the pictures."

Sample Profile Comments
- "Lucy recognizes and uses ordinal numerals and words in context."

**Matching**
- Tell students to look at the ants again. Say, "Make tally marks along the bottom of your page to show how many ants altogether."

Sample Profile Comments
- "Peter knows how to record and count using tally marks."

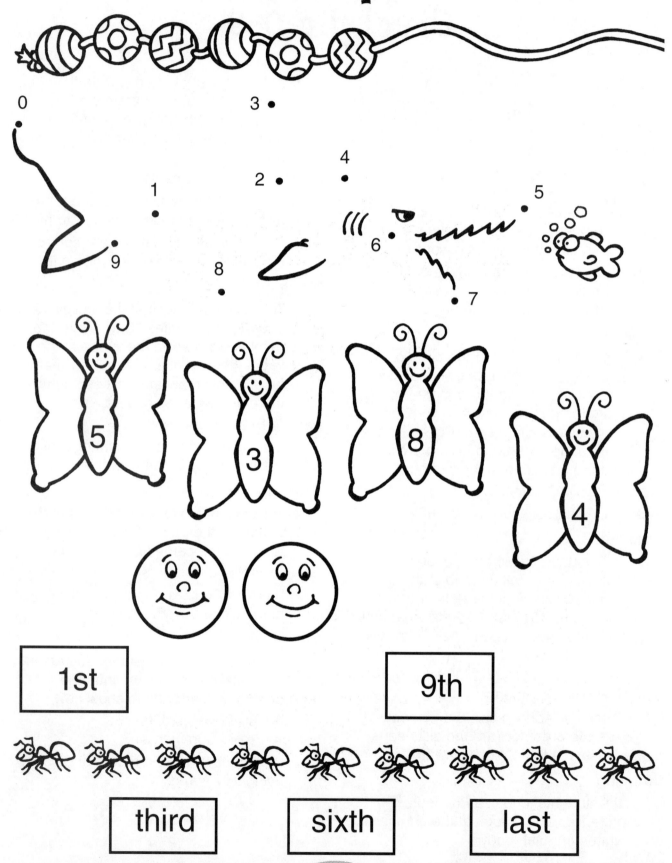

# Exploring 10–90

**In this unit, your students will:**

- ❏ Count forward/backward by 10s
- ❏ Model groups of 10 ones as 1 ten
- ❏ Estimate, model, and count groups of 10
- ❏ Match and order sets of objects, numerals, and words for 10–90
- ❏ Record 10–90 as numerals or words

*Exploring 10–90*

# Activities

## Clap to 10

**Activity**

This activity is a great way to make groups of ten special for all students.

- The whole class stands in a circle.
- The pupils start clapping hands together, counting aloud "one, two, three, . . ." as they clap.
- When everyone reaches "ten," suggest everyone do a gigantic jump calling out "ten" excitedly. Then start counting all over again.
- Repeat until you think the class has passed its enjoyment level!

**Variation**

- Have students try flicking their fingers as you count by ones to "ten." Then clap their hands when you reach "ten."
- Challenge the class to invent their own actions to make counting to ten special in some way.

## Counting by 10s

**Resources**

hundreds chart

**Activity**

- Ask whether anyone knows how to count aloud by tens starting with 0.
- Ask them to explain to the class how they know which numbers to say next.
- Discuss different strategies.
- Reveal the hundreds chart. Show how the numbers on the hundreds chart all end with a multiple of ten.
- Ask everyone to close their eyes while you count aloud by tens to "ninety" or "one hundred." Try counting by tens backwards as well as forwards.
- Practice counting by tens to ninety in the following ways: forward and backward, softly and loudly, or slowly and quickly.

## Pass the Ball

**Resources**

a large ball for each group

**Activity**

- Form groups of up to ten students.
- Have each group stand in a circle at least an arm's length or so away from each other.
- Have students pass the ball (e.g., a chest pass) to the next person in a clockwise direction, everyone counting aloud by tens with each pass until you reach "ninety."
- Then try counting backward down to "ten" again as the ball passes around the circle.

**Variation**

- Pass the ball to students at random, saying the next multiple of ten before you catch the ball each time.

# Exploring 10–90

# Activities

## Skipping by 10s

### Resources
large skipping rope for each group

### Activity
- Each group elects two people to stand at each end of the rope. They are the turners.
- The rest of the team stands in a line and takes turns to skip, counting by tens from 0 to 90 aloud without getting their feet tangled. Ask, "Can anyone continue skipping while counting by tens from 0–90 then back to 0 again?"

## What a Lot of Fingers

### Activity
- Ask individual students to stand at the front of the room with their ten fingers held up in front of them. Ask, "How many groups of ten?" Practice counting by tens aloud as you point to each person. Ask, "What if the whole class stood in a circle or a line? What number would we reach?" Have students guess first and then find a way to check their guesses.
- Ask, "What's the largest multiple of ten to which you can count?"

### Variation
- The whole class sits in a circle on the playground with their shoes and socks off! Practice counting aloud by tens by counting all the groups of ten toes.
- Ask, "How many toes altogether? Did you know that some people have 12 toes and 12 fingers?"

## Reference Pictures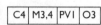

### Resources
a copy of each of the 10–90 reference pictures (pages 26–34 colored and laminated)

### Activity
A: Discuss the groups of ten in each picture. Remind the students how easy it is to count how many things altogether when you know that there are 10 objects in each group. Tell students that counting by tens means you do not have to count each item one by one.

B: Have students look at the numbers on each picture. Ask, "Can anyone see a pattern in how the numbers are made?" When counting aloud, reinforce the different names for each number (e.g., "thirty," "three tens," "three groups of ten people").

C: Place the pictures in counting order at the front of the room for ready reference throughout the year.

**Reference Pictures** Exploring 10–90

one ten
ten
10

Reference Pictures                                          Exploring 10–90

two tens
**twenty**
**20**

27

**Reference Pictures**  Exploring 10–90

four tens
**forty**
**40**

**Reference Pictures**  Exploring 10–90

50
five tens
fifty

**Reference Pictures**  Exploring 10–90

six tens
**sixty**
**60**

**Reference Pictures**        **Exploring 10–90**

seven tens
**seventy**
**70**

**Reference Pictures**    **Exploring 10–90**

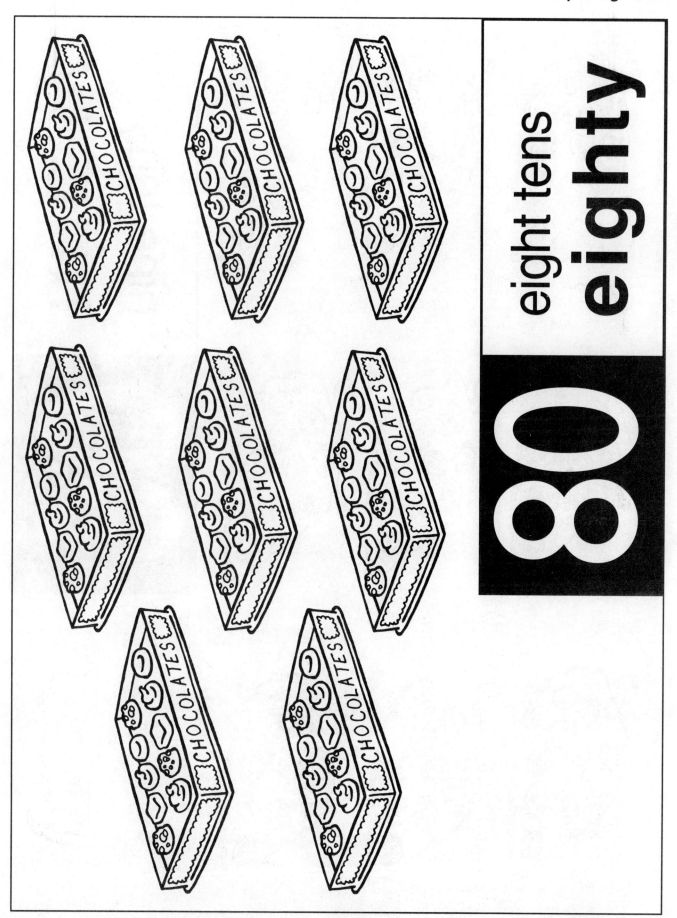

**Reference Pictures**  Exploring 10–90

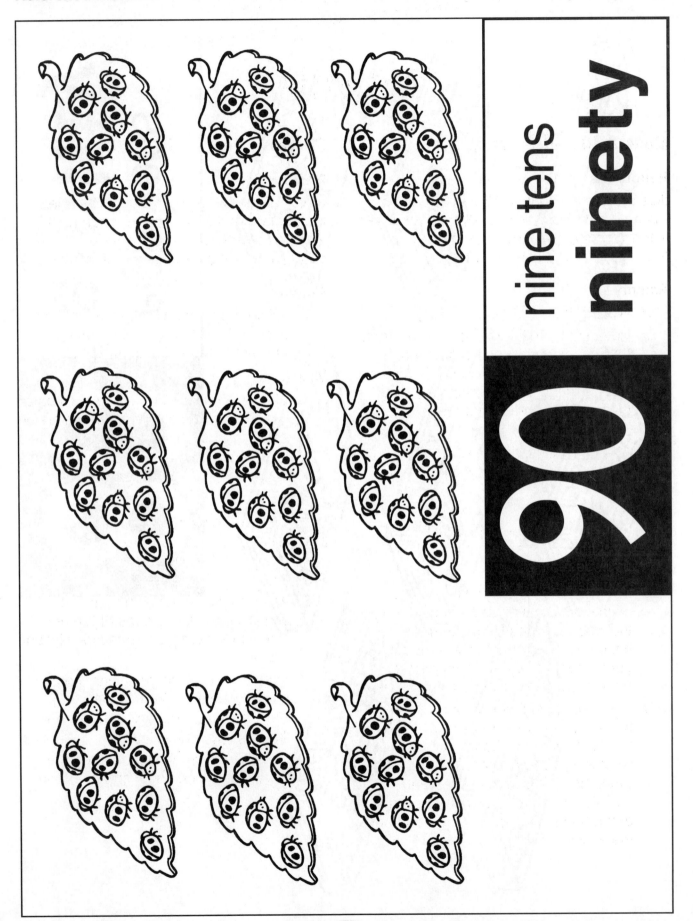

nine tens
ninety
90

# Activities

## Make It 10

### Resources
dice or spinners (pages 15 and 16), dried kidney beans, craft sticks, hobby glue, "Make It Ten Worksheet" (page 36) cut into four (one for each student), pencils

### Activity
- Work in teams of four.
- Have students take turns to throw a die (or use a spinner) and collect the matching number of beans. Once they have a group of 10 beans, they glue these along a craft stick to make a "tens stick."
- At a given time signal (e.g., after 5 minutes), have students count their total number of tens sticks. Count by 10s to discover how many beans have been collected altogether—"one ten," "two tens," "three tens," etc.

### Variation
- Use the tens sticks for whole class counting by 10s at other times during the week and year (e.g., Grab a handful of tens sticks. Have students guess how many beans altogether. Have them check by counting in tens.)
- Use a die or spinner as above, but have students record their groups of ten by drawing ten beans on each blank stick on the "Make It Ten Worksheet." Have them write the numeral which tells how many groups of ten beans altogether at the end of the activity.

## Plenty of Pasta

### Resources
dice or spinners (pages 15 and 16), dried pasta bows or shells (small enough to fit 10 along a tongue depressor), wooden tongue depressors, hobby glue

### Activity
- Same directions as "Make It Ten," but use pasta and tongue depressors instead of beans and craft sticks.

### Variation
- Thread 10 beads or pasta pieces onto bead string to make "tens" bracelets or necklaces. Challenge the students to invent their own ways to show groups of ten.

## What Else Comes in Tens?

### Activity
- Ask the class to name things that definitely come in groups of ten (e.g., sides on a decagon, cents in a 10 cent coin).
- Ask the class to name things that possibly come in groups of 10 (e.g., wheels on a large truck, petals on a flower, puppies in a litter).
- Collect all these ideas together in a large class book about groups of ten.

**Make It Ten Worksheet**  Exploring 10–90

## Make It Ten

## Make It Ten

## Make It Ten

## Make It Ten

Exploring 10–90

# Activities

## Using 10–90 Numeral/Finger/Number Word Cards

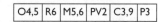

### What to Do

| O4,5 | R6 | M5,6 | PV2 | C3,9 | P3 |

- Make a copy of the 10–90 Numeral Cards (page 38), Finger Cards (pages 39 and 40), and Number Word Cards (page 41 or page 42). Laminate and then cut out as sets of 10 cards for whole class demonstrations.
- Make copies of the 10–90 Numeral Cards (page 38), Finger Cards (pages 39 and 40) and Number Word Cards (page 41 or page 42). Then cut out as sets of 10 individual cards for small group activities.

### Whole Class Ideas

- Discuss the set of large finger cards. Select cards at random and talk about what they show (e.g., How many sets are there of ten fingers altogether?) Do the same for the set of large numeral and number word cards.
- Shuffle the large cards. Select a card at random and ask two students to race each other to model the number shown with bean or pasta sticks.
- Hand each student one card. On a given signal, have students race to discover their partners by finding the matching numeral, finger, and number word cards. Repeat using different cards for each student.
- Sort the cards into the three separate sets. Shuffle the numeral cards. Ask individual students to explain aloud how to sort them into counting order forward or backward. Repeat for the finger and number word cards.
- Shuffle the cards. Have students select three at random and sort from the smallest to the largest number.
- Shuffle the cards. Have students select a starting card and count aloud from that number by tens to 90.
- Shuffle the cards. Have students select a starting card and count back from that number by tens to 0.
- Shuffle the cards. Have students select three cards. Ask, "Which one shows the largest number? The smallest? The number in between?"
- Place the ten cards in one set in order at the front of the class. Ask the students to close their eyes while you secretly turn over three cards at random. When they open their eyes, see how quickly they can name the hidden numbers.

### Small Group Ideas

- Repeat all the activities above using the smaller card sets.

0–90 Numeral Cards

Exploring 10–90

| 0 | 10 |
| 20 | 30 |
| 40 | 50 |
| 60 | 70 |
| 80 | 90 |

Finger Cards — Exploring 10–90

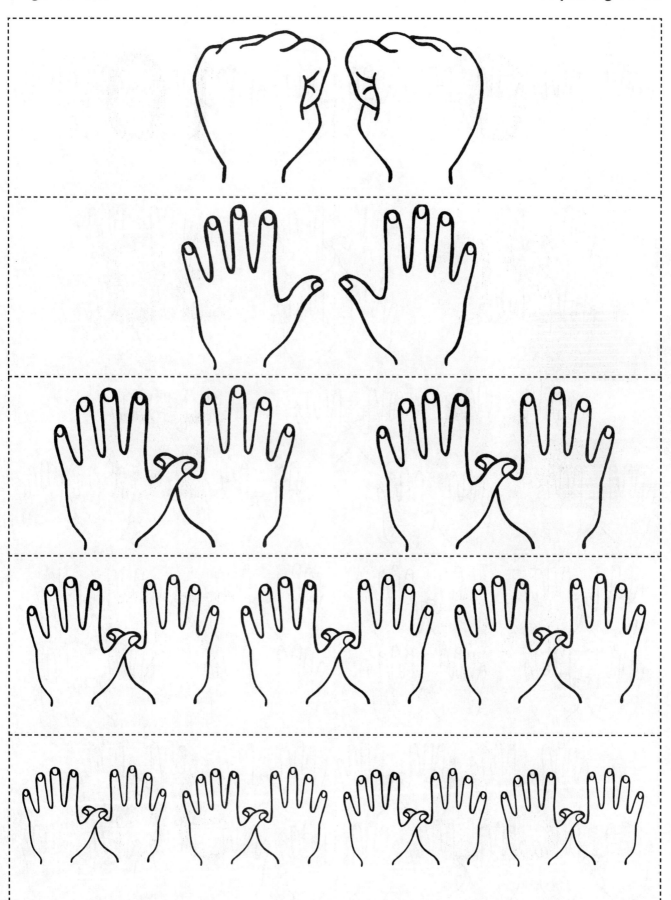

Finger Cards — Exploring 10–90

**Finger Cards**  Exploring 10–90

Finger Cards  Exploring 10–90

**Number Word Cards**          **Exploring 10–90**

| | |
|---|---|
| zero | ten |
| twenty | thirty |
| forty | fifty |
| sixty | seventy |
| eighty | ninety |

**Number Word Cards** — **Exploring 10–90**

| no tens | one ten |
| two tens | three tens |
| four tens | five tens |
| six tens | seven tens |
| eight tens | nine tens |

# Exploring 10–90

# Activities

## How Far Can You Go?

### Resources
a calculator, scrap paper/pencil for each student

### Activity
- Tell students the following directions: "Press (on). Press (10). Press (+) (+). Press (=). Write 10 at the top of your paper. Press (=) again. What number is showing now? Write this new number under the first. Guess what will happen when you press (=) again? Check your guess by pressing (=). Repeat. Continue until you cannot guess anymore. What number did you reach? Compare your answers with a friend. Are the numbers the same? Why?"

## Paper People Chain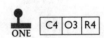

### Resources
colored paper and scissors for each student

### Activity
- Demonstrate how to make paper people by accordion folding then cutting out the outline of a person, with hands and feet attached. Have students experiment with how many folds they can make and how many people they end up with. Have students draw sweat shirts on each of the people. Then have them write in multiples of 10 in order forward or backward. Ask, "How many groups of 10 can you make?"

## Sort the Lion's Mane

### Resources
10 colored plastic pegs (each with a number from 0–90), a paper plate (with a lion's face in the center), storage container

### Activity
- Have students make the lion's mane by placing pegs around the edge of the paper plate in counting order from 0–90. Have them ask their partners to close their eyes while they rearrange two pegs. Ask, "Can your partner spot the mistake and correct it quickly?"

### Variation
Have students sort 10 different "zero" to "ninety" pegs in the same way.

## Which Path?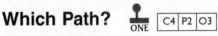

### Resources
Rabbit/Frog Worksheet (page 44), cut in two

### Activity
- Ask students to help the rabbit munch all the carrots by drawing a line from each carrot to the next, counting by 10s from 0 to 90. Then have them help the frog jump from lilypad to lilypad by drawing a line from one lilypad to the next, counting by 10s from ninety to zero.

# Activities

## That's a Lot of Students!

### Activity
One day before you go back into class—after recess or after lunch—ask all the students in your grade to guess how many groups of 10 they could make. Have them guess first and then find a way to check all the guesses (e.g., sit down in rows or in circles).

## Recording Tens

### Resources
chalkboard, chalk, "zero" to "ninety," "no tens" to "nine tens" word cards

### Activity
Draw four large shapes on the board. Form four teams.

A: Call out a number from 10–90. Teams send their first members to write this number as a word in the middle of their shapes. Repeat until everyone has had a turn.

B: Hold up a "zero" to "ninety" word card (e.g., forty). Teams race to record this on the board as a number (e.g., 40). Repeat until everyone has had a turn.

C: Place the cards "no ten" to "nine tens" face down. Turn over a card (e.g., three tens) and call it out. Teams race to record this on the board as a number (e.g., 30). Repeat until everyone has had a turn.

D: Repeat the game above, but this time hold up the card without saying it aloud. Teams race to write it as a number on the board.

E: Hold up a number card from 10–90. Teams race each other to record this on the board as a number word "zero" to "ninety." Repeat until everyone has had a turn.

## How Many Tens?

### Resources
"How Many Tens?" (page 46)

### Activity
Discuss the bears in general. Have students guess how many groups of ten bears there are altogether. Have them find a way to check the guess (e.g., by drawing a line around groups of ten bears; coloring in ten bears yellow, then ten bears green).

### Variation
Have students find another way to use this worksheet. Have them look through the whole scene for bears that are alike in some way (e.g., bears holding an ice-cream, bears with bow ties). How many bears are in each category?

**Exploring 10–90**

# How Many Tens?

46

# Exploring 10–90

# Activities

## Mental Mind Munchers  C4,5 | P2,3 | O3 | M3

Try a selection of the following types of oral questions for about three minutes daily.

- ❑ "I am counting groups of 10 people at a party. 40, 50, . . . . What is the next number in my pattern?"
- ❑ "I am counting backwards by groups of 10. 90, 80, 70, . . . . What is the next number in my pattern?"
- ❑ "I am counting by 10s but I am going to skip a number. Can you discover it? Ready? 30, 40, 50, 60, 80, 90. What number did I miss?"
- ❑ "This time I am counting by 10s backwards. I am going to skip a number. Ready? 90, 80, 70, 50, 40, 30, 20. What number did I skip?"
- ❑ "What is my number? I am thinking of 'a group of tens' number that is smaller than 40. What could it be?" (e.g., 20, 30)
- ❑ (Show up to nine beansticks for a few seconds and then hide them.) "How many beans did you see? Were there more than 50? Less than 80?" (e.g., 6 beansticks—that is 60 beans altogether)

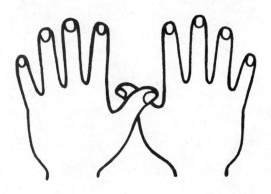

## Check-Up 10–90   M3,4 | R4 | PV1 | C4 | O3

### Resources
"10–90 Check-Up" (page 48)

### Activity
Use this worksheet with small groups or with the whole class as a written form of assessment. Record students' responses on the Skills Record Sheet (page 94). The following are suggested instructions.

### Counting, Matching, Recording, Place Value

"Look at the numbers and number words at the top of the page. Join the ones that belong together by drawing a line."

"Look at the beetles. Write the number which shows you how many spots on all the beetles together."

"Look at the bundles of sticks. Draw a circle around some of the bundles to show thirty sticks altogether."

"Look at the boxes. Draw 10 apples in each box. Write the number which tells you how many apples altogether."

"Look at the shirts with numbers in the middle. Look at the shirts without any numbers. Write the missing numbers when you count by 10s."

### Ordering

Ask students to find the "join the dots" picture. Ask them to guess what it might be. Say, "Start at 90. Join the dots counting backwards by 10s to zero. What do you think your picture is now?" Have them color it to make an interesting pattern.

# Check-Up 10-90

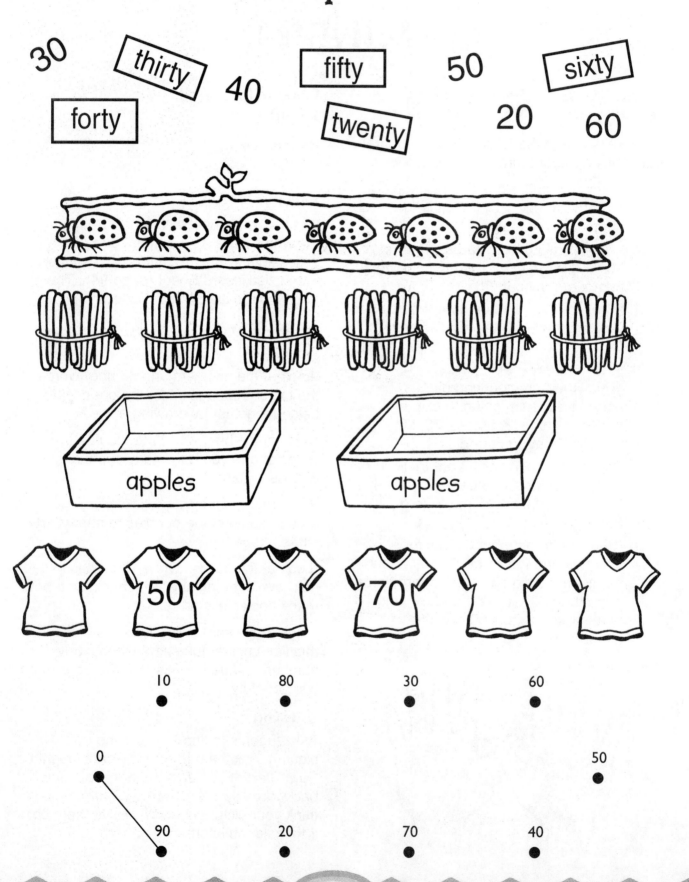

# Exploring 11–19

**In this unit, your students will:**

- ❏ Explore number patterns to 20
- ❏ Identify one more or one less than a given number to 20
- ❏ Identify two more or two less than a given number to 20
- ❏ Estimate, model, and count 11–20 objects
- ❏ Model numbers 11–20 as one group of 10 and extras
- ❏ Match and order sets of objects, numerals, and words for 0–20
- ❏ Record 0–20 as numerals, words, or tally marks

# Activities

## What's So Special About 11–19?

**Resources**

Reference Pictures for 11–19 (pages 51–59), Numeral Cards/Caterpillar Dot Cards/Number Word Cards (pages 60–62)

**Activity**

- Ask 11 students to stand at the front. Ask, "How many groups of ten can you make?" Discuss the fact that you can make one group of ten and one extra. Ask, "What should we do with the extra person?" Discuss different suggestions. The extra person is the start of a new group of ten.

- Have students look at the reference picture for 11 on page 51. Have them look at the group of ten people and one extra. Have them match the digit and number word cards to the picture. Reinforce the idea that there is one ten and one extra.

**Variation**

Show the reference picture for 12 on page 52. Ask individual students to explain how many objects they see (e.g., one group of 10 and two extras). Repeat for all the number reference pictures to 19. Have students find the matching numeral/caterpillar/word cards for each picture.

## Making Patterns

**Resources**

counters, 11–19 numeral cards (page 60) for each group.

**Activity**

- Have students shuffle the cards and place them face down. Each person turns over a card.

- Have students take a matching number of counters and explore all the ways they can make patterns with that many counters.

- Have students compare their discoveries with other people in their group. (e.g., 13)

**Variation**

- Have students record some of their discoveries on sheets of paper.

## Guess and Check

**Resources**

19 counters, two sets of digit cards 0–9 (page 11), Number Word Cards (page 12 and page 62) for each pair

**Activity**

- Have students work with a partner and place some counters in their hands while their partners look away.

- Have them ask their partners to look at the counters and guess how many they have.

- Have students check by counting together. How many groups of 10? How many extras?

- Have students exchange roles. Ask, "Do your guesses get closer with practice?"

**Variation**

Have students match digit cards and the number word to each set.

**Reference Pictures**  Exploring 11–19

1 ten   1 one

**Reference Pictures**   **Exploring 11–19**

## 1 ten    2 ones

## 12    twelve

**Reference Pictures**  Exploring 11–19

1 ten    3 ones

| 13 | thirteen |

**Reference Pictures**  Exploring 11–19

## 1 ten    4 ones

# 14 fourteen

**Reference Pictures**  Exploring 11–19

1 ten    5 ones

15    fifteen

**Reference Pictures**  Exploring 11–19

# 1 ten  6 ones

# 16  sixteen

**Reference Pictures**  Exploring 11–19

1 ten    7 ones

# 17 seventeen

**Reference Pictures**  Exploring 11–19

1 ten    8 ones

# 18 eighteen

**Reference Pictures**  Exploring 11–19

1 ten    9 ones

# 19 nineteen

**Numeral Cards**                                           Exploring 11–19

| 11 | 12 |
| 13 | 14 |
| 15 | 16 |
| 17 | 18 |
| 19 |    |

**Numeral Cards**

**Caterpillar Dot Cards**  Exploring 11–19

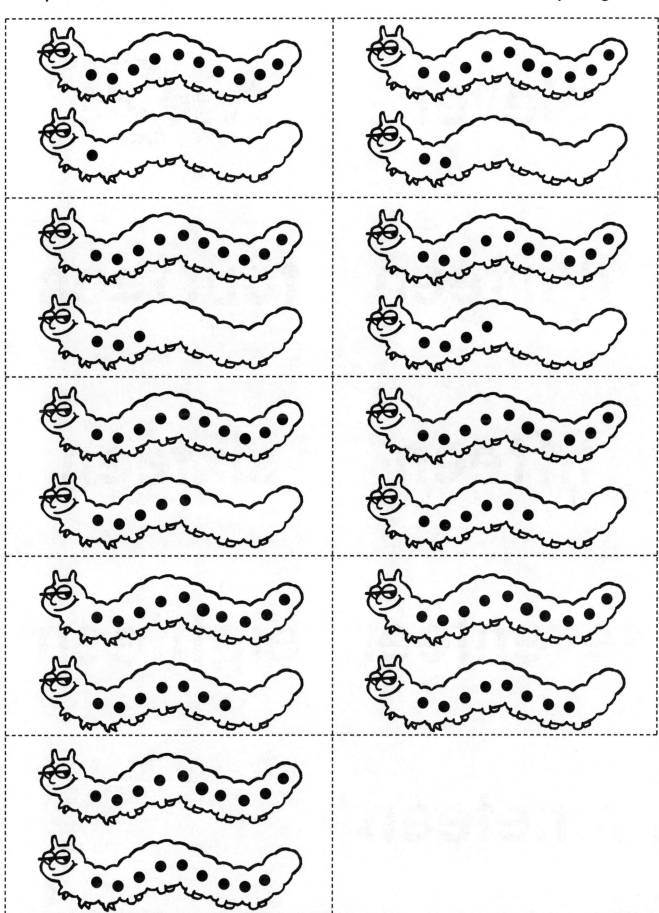

**Number Word Cards**  Exploring 11–19

| eleven | twelve |
| --- | --- |
| thirteen | fourteen |
| fifteen | sixteen |
| seventeen | eighteen |
| nineteen | |

Exploring 11–19

# Activities

## That's a Lot of Jumps

**Resources**

balls, jump ropes, beanbags, hoops, stopwatch

**Activity**

- Suggest activities that everyone can do together (outside or in a hall) involving the numbers between 10 and 20 (e.g., 13 jumps, 15 ball passes, run to the fence and back 19 times, toss a beanbag to a partner 14 times, roll a hoop for 17 seconds using the stopwatch).

**Variation**

- Use the stopwatch to discover the maximum number of events in a given time limit from 10–20 seconds (e.g., In 15 seconds, Joe can bounce the ball 12 times.). Record in a class record book.

## Let's Make Music

**Resources**

variety of musical instruments (enough for everyone in the class to have an instrument)

**Activity**

- Same directions as "That's a Lot of Jumps" but with a musical twist! Suggest you invent special "10–20 music." Ask for suggestions for activities that everyone will do together involving the numbers between 10 and 20 (e.g., 16 taps on a triangle followed by 11 drumbeats).

**Variation**

- Perform musical items based on 1–20 at the school assembly.

## How Many Are There?

**Resources**

"How Many Are There?" (page 64)

**Activity**

- Ask students to look at each picture. Together decide on 11–19 things to draw to complete each picture. Have them write the matching number and number word beside each picture, too. (e.g., fishbowl—draw 14 fish, dog—draw 15 spots, cake—draw 18 candles, smile—draw 13 teeth)

- Ask students to turn over their pages and create their own activities to do with counting from 11–19.

**Variation**

- Use this worksheet again with a different set of numbers from 11–19. It can also be used for numbers from 1–10 if some students find the larger numbers difficult to match with objects.

**Exploring 11–19**

# How Many Are There?

# Activities

## Bundling 10s and Extras

### Resources
beansticks from "Make It Ten" (page 35), extra dried beans, 0–9 spinner (pages 15 and 16), counters for each group, two sets of 0–9 digit cards, pencil/paper for each student

### Activity
- Have everyone start with a 10s stick. In turn, spin to see how many extras you will take. Collect the matching number of beans. Tell everyone how many beans you have altogether (e.g., "I have one tens stick and three extra beans—that's thirteen."). Match this number with the digit cards. The person who makes the largest number for each round collects a counter. Return the loose beans to the center before starting the next round.

### Variations
- Record your score on paper (e.g., by tallying, drawing a tens stick/extra beans, writing the numeral). Remind the students that you would only make a new tens stick if you collect ten extra beans).
- Use pasta/pasta sticks (page 35), craft sticks/elastic bands, or toothpicks/elastic bands to repeat this activity.
- Use the Caterpillar Dot Cards (page 61). Shuffle. Have students turn over the top card and race to model the number using beans/pasta/craft sticks/toothpicks.

## Hit the Target

### Resources
a calculator for each pair

### Activity
- Have students press (on). Then have them press (1) (0). Instruct them to press (+) (+) (1). Before they press the next button, have them identify a target number between 11 and 19. Then have them guess how many times you have to press the (=) button to make it display the target number. Have them check by pressing (=). Were they right?

## How Many Can You Find?

### Resources
"How Many Can You Find?" (page 66), pencil, paper

### Activity
- Review how to record 1–10 as tally marks. Demonstrate how to tally from 11–19 events. Discuss the "How Many Can You Find?" picture. Ask, "Can you find all the objects that are the same? Draw a picture to show what you are tallying. As you find an example, color it (or cross it out) and make a tally mark beside your picture. Count up your tallies. Which object appears the most? The least? Which is your favorite object? Why?"

**Exploring 11–19**

# How Many Can You Find?

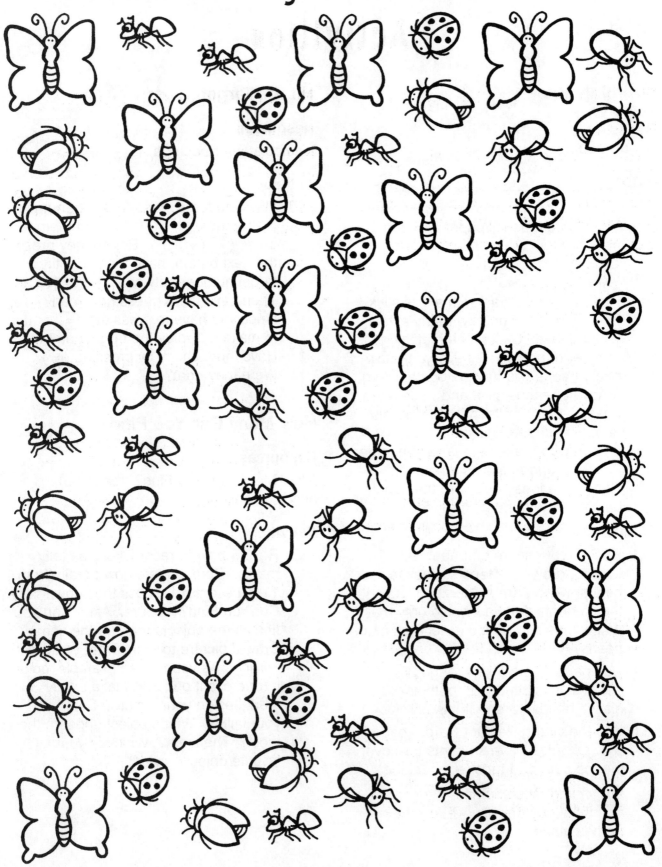

# Activities

## Find a Number

**Resources**

paper to make a large class book

**Activity**

- ❏ Identify as many real-life uses for the numbers 0–20 as you can.
- ❏ Put the ideas together into a large class book.

   e.g., "There were 12 candies in my packet."

   "Today is the thirteenth of May."

   "My house is the eighteenth on our street."

   "Dad works on the nineteenth floor of his building."

   "Grandma gave me $15 dollars for my birthday."

   "Our dog is fourteen years old."

   "There are sixteen boys at school today."

## Back to Front

**Activity**

- ❏ Point out that when people say the numbers from 11–19, they sometimes hear something strange (e.g., 14). People say "fourteen," yet they write it as one group of ten and four extras—not the other way around! Have students find some other numbers like this (e.g., seventeen). Say, "When you are writing these numbers, try not to get tricked by the way we say their names aloud."

## Number Roll-up x 2

**Resources**

Number Roll-up (Cut page 68 into vertical lengths along the unbroken lines. Paste the ends together to make a long strip. Fold up along the horizontal dotted lines, starting from the bottom.)

**Activity**

- ❏ Have students look at each bicycle on their number roll-ups. Ask, "How many wheels can you see?"
- ❏ Have students count the wheels aloud by 2s, unfolding the number strip as they count. Ask, "How fast can you count and unroll? Can you count backward by 2s as well?"
- ❏ Have students open their number strips and count backward from 24 by 2s.
- ❏ Have students open at any place on their Number Roll-ups. Ask, "Can you start counting forward or backward from that number?"

**Variation**

- ❏ Have students make their own Number Roll-ups by folding a long strip of paper into sections. Have them open it again and draw their own counting pictures. (e.g., To count by 3s: Draw triangles. Count the number of sides. To count by 4s: Draw dogs. Count the number of legs. To count by 5s: Draw an outstretched hand. Count fingers.)

Exploring 11–19

# Number Roll-up x 2

Paste bottom of first strip here.

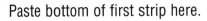

Paste top of third strip here.

Cut off this section.

# Activities

## Please Sit Down

### Activity
- Have everyone stand up. Call out a number from 1–20. The class has to cooperate to finish with the matching number of students standing and the rest sitting down.

## Frog Jumps

### Resources
20 beanbags (or 20 chalk dots on the ground) 20 inches or 50 cm apart

### Activity
- In the playground, tell the story of how frogs love to jump from lily pad to lily pad. Have students pretend each beanbag is a lily pad.
- Discuss the starting point, where the feet are together before they jump to the first lily pad. Call this "Start." Ask, "Can you jump like a frog from the start to the last lily pad counting each jump aloud?"

### Variations
- Invent stories about a jumping frog. Select individual students to model each story as you tell it (e.g., This frog jumped along 12 lily pads. Then she had a rest. She then jumped along 3 more lily pads. How many jumps were there altogether?).
- Create stories that involve jumping backward as well as forward. Start at any number.
- Label each "lily pad" by writing "Start" in chalk, and then a chalk number from 1 to 20 beside each beanbag. This activity will prepare you for working with number lines later in the year.
- Have students jump over two lily pads at a time and count by 2s from 2 to 20 forwards or backwards.
- Place the lily pads closer together. Have students jump by 5s to 20 forwards and backwards.
- Use just 9 lily pads. Review counting by 10s from 10 to 90 by writing 10, 20, . . . 90 in chalk beside each lily pad. Have students start at any lily pad and jump forwards or backwards.

## Join the Dots

### Resources
Join Them Up (page 70), cut out as two separate activities

### Activity
- Have students guess what each picture will look like when completed.
- Have students join the dots in counting order forward or backward.
- Ask students, "Was your guess close?"

### Variation
- Have students invent their own 1–20 dot picture clues for a friend to use.

**Exploring 11–19**

# Join the Dots

# Join the Dots

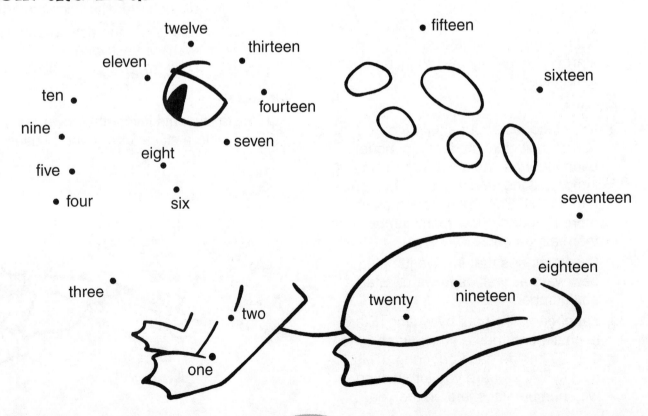

Exploring 11–19

# Activities

## All Stations to Central

### Resources
scrap paper and a pencil for each player, counters

### Activity
- ❑ Explain how trains pass through stations in order along a railway track.
- ❑ For this game, have students pretend that numbers are stations and that the line you draw is the railway track. "Central" is the station at the end of the line.
- ❑ Have students work in groups.
- ❑ Each person writes the numbers 1–20 at random on his or her paper. These are the "Stations."
- ❑ Have students exchange papers and, on a signal, join the random numbers into counting order from 1–20. The first person to finish calls out "Central." The rest of the group checks their lines and the winner collects a counter.
- ❑ Repeat.
- ❑ At the end of the activity (e.g., after 5 minutes), count to see who has the most counters.

### Variations
- ❑ For an easier game, have students write the numbers from 1–10.
- ❑ Have students join the numbers in descending order.
- ❑ Have students write only even or odd numbers.
- ❑ Have students write multiples of 5 (to 50) or 10 (to 100 or beyond).
- ❑ Have students write ten random numbers from 0–50.

## Paw Prints

### Resources
0–20 Paw Prints on page 72 (e.g., copied on green paper), laminated and cut out as individual paw prints

### Activity
- ❑ Tell a story about a bear walking through mud or snow, leaving all the paw prints behind him as he goes.
- ❑ Have students sort the paw prints into counting order from 1–20, or backward from 20–1.

### Variations
- ❑ Have students sort the paws in to odd or even numbers, forward or backward.
- ❑ Have students sort the paw prints into counting order. Then have them turn some face down and ask the partner to identify the missing pieces.
- ❑ Have students take any paw print. Then have them find the number before and the number after. Have them find the number that comes two before or two after.

**Exploring 11–19**

# 0-20 Paw Prints

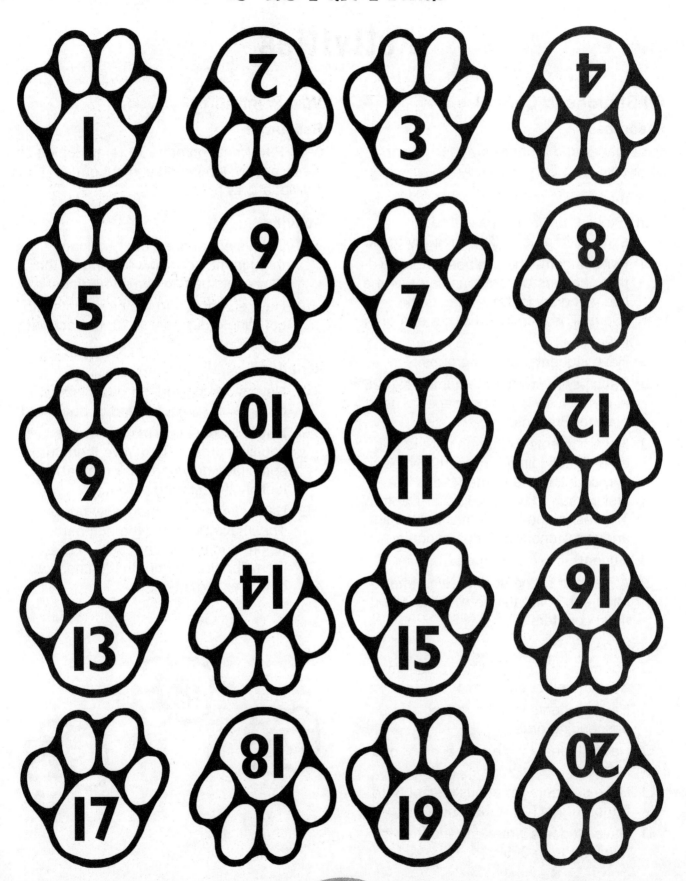

# Activities

## Who Goes Where?  P3,4 O4 C8,9

**Resources**

two copies of 10 children cards (page 18) for each pair (colored, labelled from 1–20 and cut out as 20 cards), plastic counters

**Activities**

- ❑ Have students mix up all the cards in the center. On a given signal, have them race to sort them into counting order forward or backward.
- ❑ Have students sort them into odd and even piles. Then sort them into counting order within each pile.
- ❑ Have students sort them into counting order. Have them ask their partners to have closed eyes while they rearrange some of the people. Can the partner then identify and correct the changes?
- ❑ Have students turn the cards face down. Then have them turn over a card each. The student with the highest number collects a counter.
- ❑ Have students take a card at random. Have them find the number before and after this number. Then have them find the numbers that are two before and two after this number.
- ❑ Have students invent their own activities to do with sorting numbers up to 20.

## Which Position?  O5

**Resources**

11th–20th Cards (page 74), Eleventh–Twentieth Cards (page 75), 1st–10th Cards (page 19), 20 mixed objects (e.g., toys), 20 children cards from previous activity

**Activity**

- ❑ Line up the 20 objects at the front of the room so that everyone can see them. Discuss and match the word and numeral cards together. Shuffle the "1st–20th" cards. Individual students come to the front, select a card at random, and identify the matching object in the line. Repeat using the "eleventh–twentieth" cards. Repeat this activity regularly.

**Variations**

- ❑ Individual students point to one of the objects lined up and ask another student to find the matching ordinal number or word card.
- ❑ Match the ordinal word (e.g., twelfth) and ordinal number (e.g., 12th) cards to the children cards.
- ❑ Challenge the students to invent their own ordinal matching activities.
- ❑ Repeat the whole class activities in small groups.

| 11th | 12th |
|------|------|
| 13th | 14th |
| 15th | 16th |
| 17th | 18th |
| 19th | 20th |

*Eleventh–Twentieth Cards* — Exploring 11–19

| eleventh | twelveth |
| thirteenth | fourteenth |
| fifteenth | sixteenth |
| seventeenth | eighteenth |
| nineteenth | twentieth |

*Exploring 11–19*

# Activities

## Roll-a-Snake

**Resources**

Roll-a-Snake boards (page 77), Roll-a-Snake number strips (page 78—Set 1 = numbers from 5–20/Set 2 = numbers from 21–36), scissors, glue

**Activity**

- Have students cut along the dashed lines.
- Have students thread the three number strips through the snake's body and glue the ends together.
- Have students find three numbers in counting order by rolling the numbers through to the one you want (e.g., 15, 16, 17).
- Have students try finding the numbers that are two more or less than a given middle number. Have them make up their own variations.

## Mental Mind Munchers

Here are more examples of questions to encourage fast oral responses and the development of mental number images.

- "How many numbers are there between 9 and 14?" (4: 10, 11, 12, 13)
- "Tell me three numbers smaller than 17?" (e.g., 10, 15, 12)
- "Count to 30, but starting from 14." (14, 15, . . . 30)
- (Briefly hold up a caterpillar dot card and then hide it.) "Lucy, how many spots did you see?"
- (Hold up an 11–19 card.) "Isaac, give me this many counters."
- "I am thinking of a number. It is larger than ten but smaller than 15. What could it be?" (11, 12, 13, 14)
- "Ned, count by 2s starting from 0 and finishing at 16." "Tom, count by 2s but starting from 3 and finishing at 19."
- "I am thinking of a number. It has one group of ten and seven extras. What is my number?"
- Ask the students to count silently while you clap. Stop at any time and ask individual students to identify the counting number at that point.
- "What is the sixteenth letter of the alphabet?" (P)
- "Is 13 an odd or an even number? Tell me how you know this."
- "Look at this number (e.g., 13). How do you know how many beansticks and extras this would be?"

Roll-a-Snake Boards  Exploring 11–19

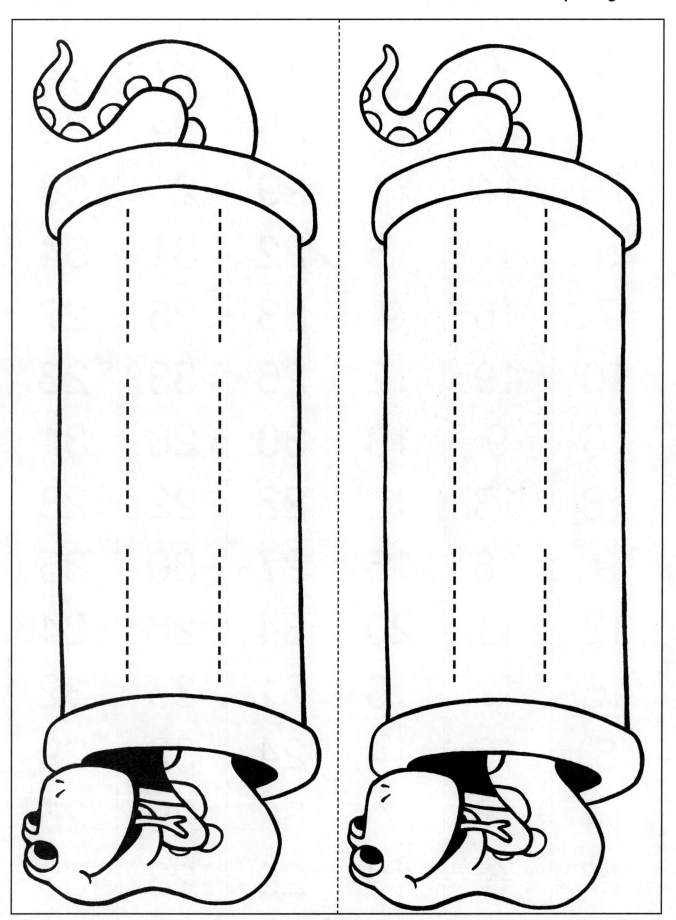

**Roll-a-Snake Number Strips**  Exploring 11–19

### Set 1 | Set 2

| Set 1 | | | Set 2 | | |
|---|---|---|---|---|---|
| 6 | 7 | 8 | 21 | 27 | 30 |
| 11 | 12 | 17 | 25 | 34 | 26 |
| 16 | 14 | 11 | 29 | 24 | 36 |
| 5 | 8 | 16 | 22 | 31 | 34 |
| 7 | 10 | 9 | 33 | 25 | 27 |
| 10 | 19 | 12 | 26 | 33 | 23 |
| 13 | 9 | 18 | 30 | 28 | 31 |
| 18 | 13 | 10 | 23 | 22 | 28 |
| 9 | 18 | 15 | 27 | 30 | 35 |
| 12 | 11 | 20 | 34 | 26 | 24 |
| 15 | 17 | 13 | 31 | 35 | 32 |
| 8 | 15 | 14 | 24 | 32 | 29 |
| 14 | 16 | 7 | 32 | 29 | 25 |
| 17 | 6 | 19 | 28 | 23 | 33 |

Exploring 11–19

# Activities

## Check-Up 11–19

**Resource** | O4,5 | R6 | M5,6 | PV2 | C3,9 | P3 |

11–19 Check-Up (page 80)

### Activity

Use the worksheet with small groups or with the whole class as a written form of assessment. Record students' responses on the Skills Record Sheet (page 94). The following are suggested instructions and sample teacher's comments used when developing student profiles.

### Ordering

- Look at the lily pads around the edge of the page. Pretend you are a frog. Find a lily pad at which to start and write the numbers from 1 to 20 in order as you jump along them. Color the fifteenth lily pad green.

Sample Profile Comments:

- "Ali knows how to write and position each of the numbers from 1–20. He has now mastered writing 3s without reversing them."

### Matching

- Look at the tiny ladybugs. Count them and color the tag which shows you how many ladybugs altogether.

Sample Profile Comments:

- "Sally has a confident grasp of everything to do with numbers to 20."
- "Although Sam circled the wrong tag, he was very close. He is constructing a clearer picture of larger numbers with daily practice."

### Place Value

- Find the beansticks. Draw how you would show 13 beans. (Ideally, the three extra beans should be drawn loose, not drawn on the second tens stick.)

Sample Profile Comments:

- "Josh understands that numbers are made up of groups of ten and extras."
- "Nina can count objects to 20 but still cannot see that they can be grouped into tens and extras. We are working on it!"

### Counting

- Look at the sets of three triangles. Write the number that comes two before each number on the left. Write the number that comes two after each number on the right.

Sample Profile Comments:

- "Ravi realizes that numbers are ordered forward or backward and can respond quickly to simple ordering problems."
- "Shannon is confident when counting forward by 1s. She still needs assistance when counting by 2s."

### Counting, Ordering

- Find the stars. Color every fifth star blue. How many stars are not blue? Write the number that tells you this. Color the seventeenth star yellow.
- Color the twelfth star red.

Sample Profile Comments:

- "Marika can count by 5s to 20 and beyond. She responds quickly to oral instructions involving ordinal numbers."

©Teacher Created Resources, Inc.    #3525 Math in Action

# Check-Up 11–19

Exploring 11–19

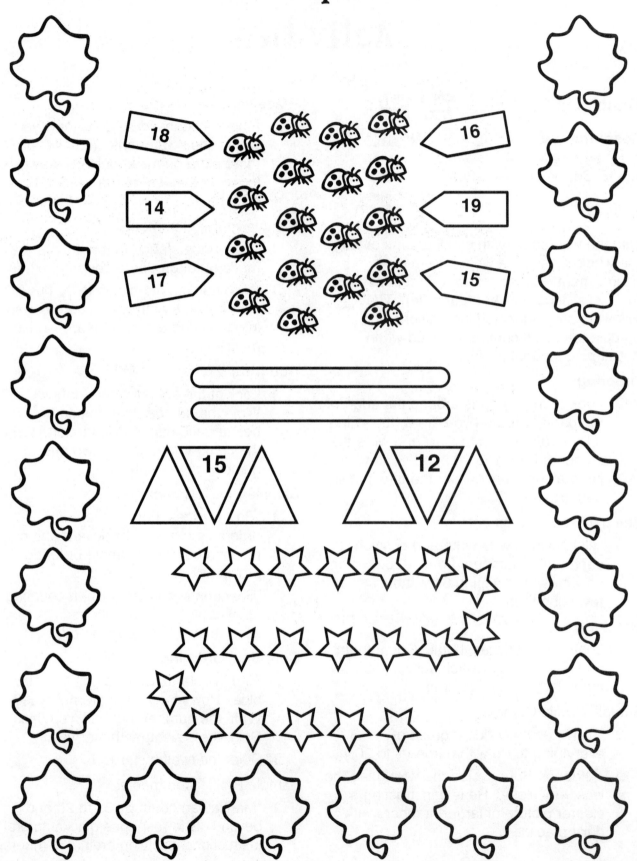

# Exploring 0–50 and Beyond

**In this unit, your students will:**

- ❏ Explore number patterns to 50
- ❏ Identify one more or one less than a given number to 50
- ❏ Identify two more or two less than a given number to 50
- ❏ Estimate, model, count 21–50 objects
- ❏ Model numbers 1–50 as groups of 10 and extras
- ❏ Match and order sets of objects, numerals, or words for 0–50
- ❏ Record 0–50 as numerals, words, or tally marks

# Activities

## Count With Me

### Activity

- Have students count aloud from 1–50 forward and backward.
- Time individual students counting as fast as they can without stumbling. Ask, "Can anyone count to 50 in less than two minutes? In less than one minute?"

### Variations

- Have students stand or sit in a circle. Have students, throughout the class, say numbers in turn as fast as they can. When they reach 50, have them start counting backward around the group again.
- Have students count by 2s, 5s, or 10s to 50 or more forward or backward.
- Have students count by 1s, 2s, 5s, or 10s from a given starting number to a given finishing number.

## What's Missing?

### Resources

copy of a 50s chart (page 83 or create a transparency)

### Activities

- Discuss the chart with the whole class. Discuss what comes before or after a given number. Secretly cover up some numbers. Ask the students to guess the hidden numbers. "What is one more or less than a given number? Two more or less? 10 more or less? What are other questions related to the chart?"

## What's My Pattern?

### Resources

transparency of the 50s chart and a separate transparency of the blank 50s chart, overhead transparency pens, pencils, copy of the blank chart for each student

### Activity

- Show the first transparency. Ask, "What patterns can you see?" (e.g., the second row numbers all start with a 1, the fifth column numbers all end with a 5). Ask, "Can you find the columns of odd numbers? Even numbers? What do you notice when you count by 5s starting from 5?" (e.g., the numbers are in two columns).
- Fill in some numbers at random on the blank chart. Ask individual students to fill in other numbers, too. Discuss what comes before or after a given number.

### Variation

- Each student fills in the numbers from 1–50 onto their blank charts.
- Experiment with different ways to record these.
  (e.g., from 1–10 across, then 11–20 . . . from 1–5 down, then 6–10 down . . . 1, 11, 21, 31, 41 down and then 2, 12, 22, 32, 42 down).

50s Chart — Exploring 0–50 and Beyond

| 1 | 2 | 3 | 4 | 5 | 6 | 7 | 8 | 9 | 10 |
|---|---|---|---|---|---|---|---|---|---|
| 11 | 12 | 13 | 14 | 15 | 16 | 17 | 18 | 19 | 20 |
| 21 | 22 | 23 | 24 | 25 | 26 | 27 | 28 | 29 | 30 |
| 31 | 32 | 33 | 34 | 35 | 36 | 37 | 38 | 39 | 40 |
| 41 | 42 | 43 | 44 | 45 | 46 | 47 | 48 | 49 | 50 |

| | | | | | | | | | |
|---|---|---|---|---|---|---|---|---|---|
| | | | | | | | | | |
| | | | | | | | | | |
| | | | | | | | | | |
| | | | | | | | | | |
| | | | | | | | | | |

*Exploring 0–50 and Beyond*

# Activities

## Can You Make It?

**Resources**

place value materials (e.g., beans/beansticks, pasta/pasta sticks, toothpicks/elastic bands, beads/bead string), calculators, digit cards 0–9, 0–9 spinner (page 16), 0–5 Tens Spinner (page 85) for each group, Tens and Ones Cards (page 86), blank paper, pencils, scissors for each student

**Activity**

❏ Work in small groups with one type of material in the center of each table. In turn, twirl the two spinners and model the number shown as groups of ten and extras on your Tens and Ones Card. Students in the calculator group press the corresponding numbers (e.g., 26), and then match it with digit cards on their Tens and Ones Card, saying "twenty six—That is two tens and six ones." Exchange tables after 3–5 minutes.

**Variation**

❏ In pairs, write the numbers from 1–50 onto paper and then cut these out as individual cards. Mix all the cards face down. Turn over a card and model that number with the material.

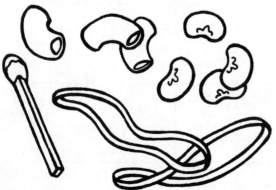

## Exchange to 50

**Resources**

beans/beansticks, dice, Tens and Ones Cards (page 86)

**Activity**

❏ Have students place the materials in the center of the group. Then have them each throw the dice and collect the matching number in beans. Tell them that when they each have ten beans, exchange them for a tens stick. Say, "Try to be the first player to have five tens sticks, or exactly 50 beans."

**Variation**

❏ Repeat in groups using pasta/pasta sticks, toothpicks/elastic bands, or beads/bead string.

❏ Play as above, but this time each player starts with five tens sticks. Have students each throw the dice and remove beans. Say, "Try to be the first player to give back all 50 beans."

## Tally It

**Resources**

blank paper, pencil for each pair

**Activity**

❏ Review how to use tally marks to record events. Identify useful events to tally within the school day (e.g., vehicles driving past the school gate, students throwing or catching a ball without dropping it). Each pair selects something to tally. Say, "Try to be the first team to tally exactly 50 points."

0–5 Tens Spinners  *Exploring 0–50 and Beyond*

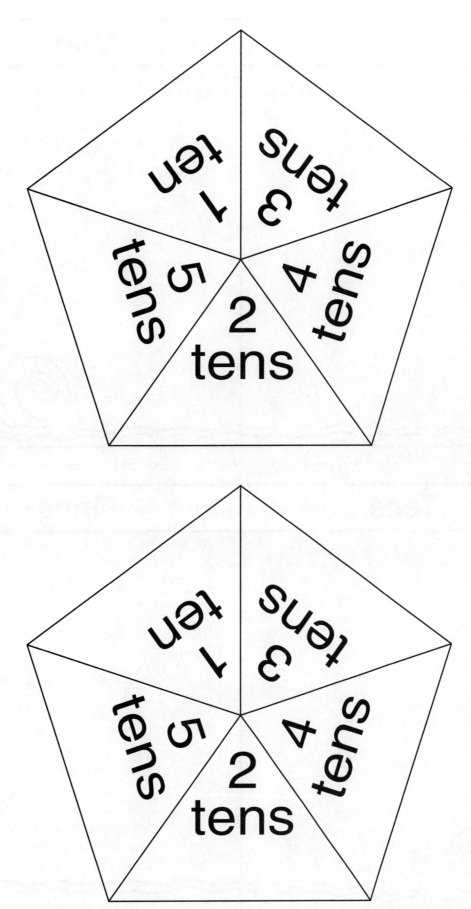

0–5 Tens Spinners

©Teacher Created Resources, Inc.  #3525 Math in Action

Tens and Ones Cards

Exploring 0–50 and Beyond

| Tens | Ones |
|------|------|
|      |      |

| Tens | Ones |
|------|------|
|      |      |

# Exploring 0–50 and Beyond

# Activities

## Guess How Many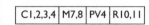

### Resources
sets of 50 (or more) counters (e.g., tiny plastic teddies), 5 small plastic counters for each group

### Activity
- Have students take a large handful. Ask, "How many are there?" Each person records his or her guesses by writing either a numeral or number word.
- Have students check by counting in different ways (e.g., by 1s, 2s, 5s, 10s).
- The player with the closest guess takes a counter.

## Show Me 50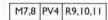

### Resources
1–5 Tens Spinner (page 85), 0–9 Spinner (page 16), scrap paper, pencils, a one-minute timer for each group

### Activity
- Have students spin the two spinners to make a number between 10 and 59. Turn over the timer. Have them find a way to record this number on paper in as many ways as they can before the timer stops (e.g., as a numeral, a number word, as tally marks, as drawings of beansticks, as groups of 10 spots and extras). Score a point for every representation.

## Croc Spots  M7,8 PV4

### Resources
Croc Spots Cards (page 88 cut into four)

### Activity
- Have students draw spots all over their crocodiles. Exchange with a friend. Say, "Guess how many spots altogether? Is there more than one way to check? (e.g., circle groups of 10 spots and count how many groups of 10)." Have them write the numbers as a numeral and as a word underneath their crocodiles. Ask, "Who has the most spots? The least?"
- Have students put their crocodiles in order.

## Paw Prints  C3,4 P4 O6 PV4 R9,10,11

### Resources
10–200 Paw Prints (page 89 e.g., blue) and 5–100 Paw Prints (page 90 e.g., red) laminated and cut out as individual paw prints

### Activity
- Have students sort into counting order forward and backward.
- Have students sort into odd or even numbers. Then have them take a card at random.
- Have students find the multiple before and the multiple after this number.

87

©Teacher Created Resources, Inc. #3525 Math in Action

**Croc Spots Cards**              **Exploring 0–50 and Beyond**

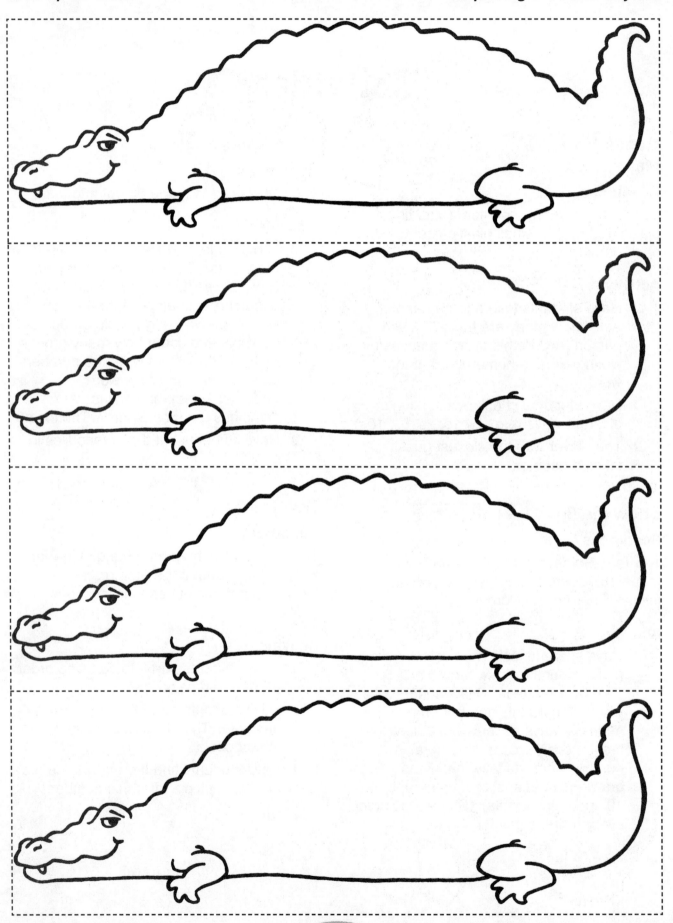

Croc Spots Cards             Exploring 0–50 and Beyond

**10–200 Paw Prints**  **Exploring 0–50 and Beyond**

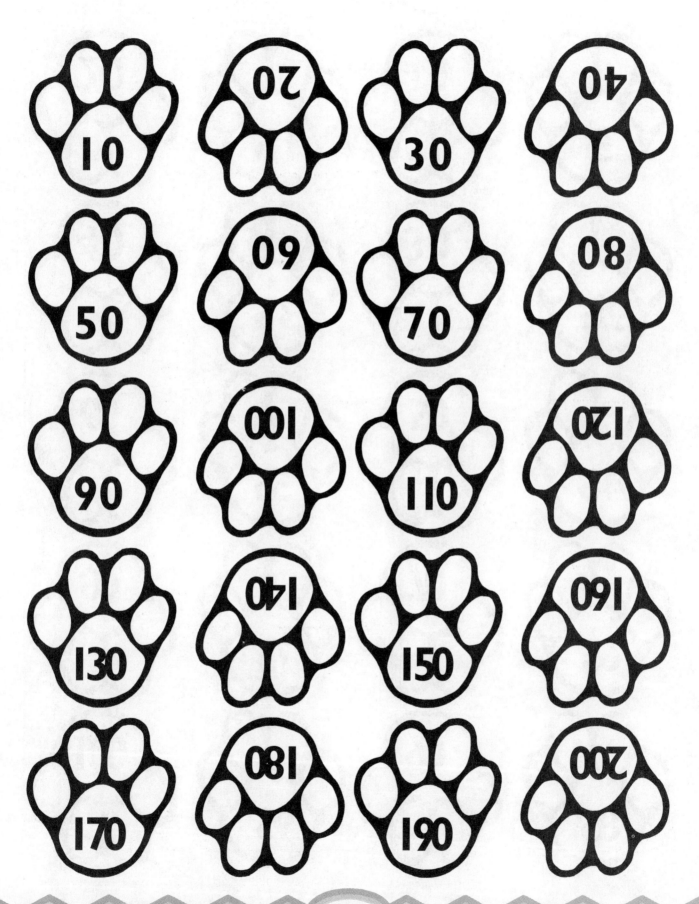

5–100 Paw Prints  Exploring 0–50 and Beyond

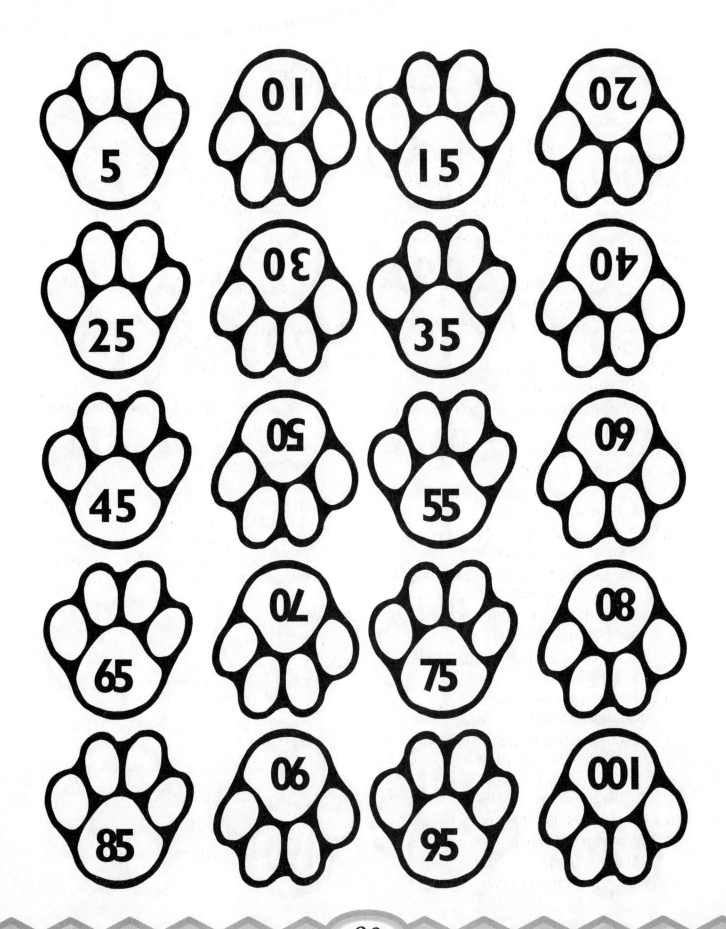

# Activities

## Mental Mind Munchers

| C2,3,4,5,10,11 | 06 | P3,4 | PV5 | R8,10 |

Find out what your students are thinking about numbers to 50 by regular questioning and discussion. Try a range of the following examples for 3–5 minutes daily with the whole class and with individuals.

- "Count forward by 2s. Start from 26. Stop when you get to 40."
- "Count backward by 2s. Start from 47. Stop at 37."
- "Count forward by 5s. Start from 25. Stop when you get to 45."
- "Count backward by 5s. Start from 40. Stop at 20."
- "Count forward by 10s. Start from 30. Stop when you get to 60."
- "Count backward by 10s. Start from 80. Stop at 50."
- "22, 24, 26, 30, 32, 34. What number did I miss in my pattern?"
- "10, 15, 20, 30, 35, 40. What number did I miss in my pattern?"
- "80, 70, 60, 40, 30. What number did I miss in my pattern?"
- "What is one more than 48? What is one less than 36?"
- "What is two more than 28? What is two less than 39?"
- "When counting by 2s, what number comes after 7? (9) After 17? (19) After 27? (29) Can you tell me the pattern?"
- "45 is 4 tens and how many extras?"
- "What is my number? It has three tens and three ones."
- "How many tens in 41?"

- "Write the number that has three tens and two ones."
- Write a number from 1–50 on the chalkboard. "Explain how you know what this number is." (e.g., 36: It has three tens and six extras. The number on the right shows how many ones. The number on the left shows how many tens. It is less than 40 but more than 30. It is even.)
- "31 and 50—Which one is the largest?"
- "43 and 12—Which one is the smallest?"
- "Tell me a number larger than 30 but smaller than 43."
- "What is the number between 31 and 33?"
- "Tell me a number between 20 and 30."
- "Write the number that is between 39 and 41."
- "Write a number that is more than 35."
- "Write the number that is just one more than 35."

*Exploring 0–50 and Beyond*

# Activities

## Check-Up 0–50

| C3 | M7,8 | O6 | PV4,5 | R10,11 |

### Resource
0–50 Check-Up (page 93)

### Activity
- Use this worksheet with small groups or with the whole class as a written form of assessment. Record students' responses on the Skills Record Sheet (page 94). The following are suggested instructions and sample teacher's comments used when developing student profiles.

### Place Value
- "Look at the boxes of crayons. Write the number that tells you how many crayons altogether. Can you write this as a word, too?"

Sample Profile Comments:

"Models numbers to 50 as bundles of 10 and extras."

### Place Value
- "Look at the next boxes. How do we write this as a number? Write it in the spaces on the right. What about the next number—46? Write how many tens and ones as words in the space on the left."

Sample Profile Comments:

"Writes numerals to 50 as words and numerals."

### Counting
- "Look at the number pattern on the caterpillar's body. Can you guess the pattern? Finish it by writing in the missing numbers."

Sample Profile Comments:

"Counts by 5s to 50 and records this as a pattern."

### Ordering
- "Look at the box with the numbers starting at 25. The naughty cat has hidden some of the numbers. Write all the missing numbers yourself."

Sample Profile Comments:

"Places numerals 0-50 into counting order."

### Place Value
- "Look at the long sausage dog. Guess how many spots altogether. Write your guess on the left. Now, put a circle around all the groups of 10 spots. Write the actual number of spots you discover at the right."

Sample Profile Comments:

"Models and counts numbers to 50 as groups of tens and ones."

### Matching
- "The fish have jumped out of their tanks, and they are all mixed up. Draw a line from each fish to its matching tank."

Sample Profile Comments:

"Matches numerals 0–50 to sets of objects."

# Check-Up 0–50

| three tens and two ones | | |
|---|---|---|
| | 4 | 6 |

5, 10, 15

| 25 | | | 28 | | | 32 | | |
|---|---|---|---|---|---|---|---|---|
| 35 | | | | 39 | | | 43 | |

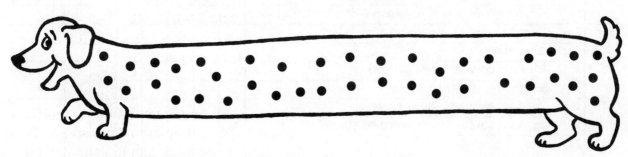

twenty to twenty-nine

thirty to thirty-nine

forty to forty-nine

# Skills Record Sheet

## NUMERATION ACTIVITIES 0–50

NAME

| | | | |
|---|---|---|---|
| Counting | C1 | Counts forward/backward by 1s to . . . | |
| | C2 | Counts forward/backward by 2s to . . . | |
| | C3 | Counts forward/backward by 5s to . . . | |
| | C4 | Counts forward/backward by 10s to . . . | |
| | C5 | Counts from one specified number to another | |
| | C6 | Identifies one more, one less than a given number to 10 | |
| | C7 | Identifies two more, two less than a given number to 10 | |
| | C8 | Identifies one more, one less than a given number to 20 | |
| | C9 | Identifies two more, two less than a given number to 20 | |
| | C10 | Identifies one more, one less than a given number to 50 | |
| | C11 | Identifies two more, two less than a given number to 50 | |
| Patterning | P1 | Recognizes, describes, and creates number patterns | |
| | P2 | Predicts and continues the next few items in a number pattern | |
| | P3 | Identifies missing items in a number pattern | |
| | P4 | Identifies a group of objects as odd or even | |
| Matching | M1 | Estimates, models, and counts 0–9 objects | |
| | M2 | Matches numerals and number words 0–9 to sets of objects | |
| | M3 | Estimates, models, and counts groups of 10–90 objects | |
| | M4 | Matches numerals/number words 10–90 to sets of 10 objects | |
| | M5 | Estimates, models, and counts 0–20 objects | |
| | M6 | Matches numerals and number words 0–20 to sets of objects | |
| | M7 | Estimates, models, and counts 0–50 objects | |
| | M8 | Matches numerals and number words 0–50 to sets of objects | |
| Ordering | O1 | Places numerals, objects, and words 0–9 into counting order | |
| | O2 | Uses, reads, and orders "1st" to "10th" and "first" to "tenth" | |
| | O3 | Places numerals, objects, and words 10–90 into counting order | |
| | O4 | Places numerals, objects, and words 0–20 into counting order | |
| | O5 | Uses "1st" to "20th" to identify positions | |
| | O6 | Places numerals, objects, and words 0–50 into counting order | |
| Recording | R1 | Uses tally marks to record counting from 0–9 | |
| | R2 | Writes numerals 0–9 | |
| | R3 | Writes number words "zero" to "nine" | |
| | R4 | Writes numerals 10–90 | |
| | R5 | Writes number words "ten" to "ninety" | |
| | R6 | Uses tally marks to record counting from 0–20 | |
| | R7 | Writes numerals 0–20 | |
| | R8 | Writes number words "zero" to "twenty" | |
| | R9 | Uses tally marks to record counting from 0–50 | |
| | R10 | Writes numerals 0–50 | |
| | R11 | Writes number words "zero" to "fifty" | |
| Place Value | PV1 | Models groups of 10 ones as 1 ten | |
| | PV2 | Models numbers to 20 as one group of 10 and extras | |
| | PV3 | Explains the value of each digit in numbers 0–20 | |
| | PV4 | Models numbers to 50 as bundles of 10 and extras | |
| | PV5 | Explains the value of each digit in numbers 0–50 | |

# Sample Weekly Program

**STRAND** Number
**GRADE** 1

**SUBSTRAND** Numeration: Studying 10, 20–90
**TERM** 1 **WEEK** 7

**LANGUAGE**
- "ten ones," "one ten"
- "more than," "fewer than"
- "as many as," "not enough"
- "There are ___ altogether in this group"

## OUTCOMES
- models groups of 10 ones as 1 ten
- counts forward/backward by tens from 10–90
- matches numerals 10–90 to groups of 1–9 tens
- sorts numerals 10–90 into counting order
- states 10 less, 10 more than a given group of ten

## RESOURCES
Hundreds Chart
Reference Picture (pages 26–34)

dried beans, craft sticks, pasta, tongue depressors, a large ball, calculators

0–90 Check-up (page 48)
10–90 Numeral/Finger/Number Word Cards (pages 38–42)

Rabbit/Frog (page 44)

3 jump ropes, colored paper, scissors plastic pegs, paper plates

| MONDAY | TUESDAY | WEDNESDAY | THURSDAY | FRIDAY |
|---|---|---|---|---|
| • Whole class game "Clap to 10" <br> • "Counting by 10s" (with Hundreds Chart) <br> • Picture reference discussion 10–90 (match numeral/number word cards to each picture) <br> • Activity: "Make it 10" "Plenty of Pasta" <br> • Finish with discussion about how easy it is to count by tens. | • Whole class game "Pass the Ball" <br> • "What a Lot of Fingers" <br> • Picture reference discussion 10–90 <br> • Discussion: "What Else Comes in Tens?" <br> • Activity: "How Far Can You Go?" <br> • Finish with whole class discussion of discoveries. | • "What a Lot of Children" <br> • "Skipping by 10s" (in 3 groups) <br> • Activities: "10 to 90 Numeral/Dot/Word Cards" <br> • Whole class challenge to finish—cover number clues on each reference picture. Guess how many then check by counting in tens. | • "Counting by 10s" (with Hundreds Chart) <br> • Rotating activities: "Paper People Chain," "Sort the Lion's Mane," "Which Path?" <br> • Homework: "How Many Tens?" worksheet | • Whole class game "Clap to 10" <br> • General discussion of homework results <br> • Review general issues <br> • Assessment task: "0–90 Check-Up" <br> • Whole class game "Pass the Ball" |

©Teacher Created Resources, Inc.      95      #3525 Math in Action

# Blank Weekly Program

**STRAND**

**GRADE**

**SUBSTRAND**

**TERM**     **WEEK**

**LANGUAGE**

- - - - -

**OUTCOMES**

- - - - - -

**RESOURCES**

| MONDAY | TUESDAY | WEDNESDAY | THURSDAY | FRIDAY |
|--------|---------|-----------|----------|--------|
|        |         |           |          |        |
|        |         |           |          |        |